W9-CSF-591

Florida Exotics

FLORIDA EXOTICS

Jack Kramer

TAYLOR PUBLISHING COMPANY
Dallas, Texas

Also by Jack Kramer

The World Wildlife Fund Book of Orchids
Orchids for the South
Women of Flowers
Earthly Delights

The author wishes to express thanks to the many growers, mail order suppliers, and hobbyists who have answered questions about their exotic plants. And many thanks to my friends who allowed us to photograph their plants—there are too many to list, so here is a communal thank you. So to all—and especially to Florida—a big thank you.

I wish to express special thanks to Holly McGuire and Stacey Sexton, my editors, who had faith in this project, and who read and reread the manuscript to make it a user-friendly book for beginner and experienced gardeners alike.

Photograph on page ii by Barbara J. Coxe; photograph on page vi by Dency Kane.

Library of Congress Cataloging-in-Publication Data
Kramer, Jack, 1927–
Florida exotics / Jack Kramer.
p. cm.
Includes bibliographical references and index.
ISBN 0-87833-921-3
1. Plants, Ornamental—Florida. 2. Exotic plants—Florida.
3. Landscape gardening—Florida. I. Title.
SB405.5.F6K73 1998
635.9'09759—dc21 98-9186
CIP

Printed in the United States of America

10 9 8 7 6 5 4 3 2 1

Table of Contents

Introduction

FLORIDA: A GARDEN ADVENTURE

Florida is blessed with an ideal tropical or semitropical climate that rarely falls below 30 degrees F. The state is divided into four hardiness zones. Zones 9, 10, and 11 are especially suitable for growing some of nature's most beautiful plants, gems that add exquisite color to your garden and brighten your total landscape. Amaryllis, agaves, yuccas, bougainvillea, allamanda, gingers, and bananas are all superb in Florida landscapes, while a great selection of tropical and semitropical trees, shrubs, vines, bromeliads, and orchids add to the year-round parade of color.

While planning your garden consider the condition of the soil. Florida soil is loamy in the panhandle, sandy in the south, and limestony in the extreme southern part of the state; all these soils need the addition of topsoil, loam, or compost. Remember, too, that a long summer growing season of intense heat and drenching rain produces lush growth in plants: allow ample space for your exotics to grow and spread.

Growing exotic plants successfully in Florida or anywhere else really depends on learning from your own experience, reading good gardening reference books, selecting plants appropriate for your area, and properly caring for your plants. I mostly plant what is available at my local nursery (thus ensuring that all my plants are appropriate for my climate), but I also order many exotics from mail order suppliers (listed in back of book).

Note: Some of the plants referred to in this book may be poisonous. Exercise care in handling new or unknown varieties of plants. The publisher and author accept no responsibility for any damage or injury resulting from the use or ingestion of or contact with any plant discussed in this book.

Botanical Names

Botanical names conform to the International Code of Nomenclature. However, taxonomists occasionally change names and some specific species names are in debate. Where there is conflict, the most current nomenclature used in growers' catalogs has been used.

Planning, Planting, and Caring for Exotics

One

THE EXOTICS

The word "exotic" has always called to mind visions of the fascinating, unusual, curious, sophisticated, and even the bizarre. To gardeners in nineteenth-century England, the new plant discoveries pouring in from all corners of the far-flung Empire personified all the aura of the exotic. Each new plant created a furor among gardeners and horticulturists; orchids from South America, succulents from Africa, and other botanical curiosities from around the world caused floralmania to grip the land.

Most of these floral fantasies were sealed in glasshouses as gardeners tried to simulate the plants' native growing conditions. In most cases, these plants died because growers had no true idea of their cultural habits, temperature requirements, and nutritional needs. Yet these failures only spurred on the collectors, increasing their intense desire to raise unusual plants.

Through the intervening decades more was learned about these imported plants. They began to thrive in England, and eventually made their way to the colonies. Today many of these once-rare plants are readily available at specialty nurseries or through certain mail order suppliers. In most parts of the United States, exotics such as orchids, bromeliads, heliconias, hedychiums, and dozens of others must be grown in the protected conditions of garden rooms or greenhouses. But in Florida many of these exotics will thrive outdoors in those areas where temperatures do not drop below 40 degrees F. The Sunshine State is a welcoming home to the exotics.

Plants of Beauty

Even within the gorgeous world of exotics, some plants surpass all others for sheer loveliness. The tropical exotics, those native to warm, humid lands, are superb examples of nature's beauty. Orchids, members of the most advanced plant family we know, are endlessly fascinating, from the incomparable spider orchid *Brassia* to the dancing lady *Oncidiums*. Bromeliads from the tropical forest floor display breathtaking color and mysterious beauty. The many species of night-blooming cacti epitomize bizarre shapes and spectacular color.

Bromeliads, with their vase-shaped growth and twisted or serrated leaves and vivid flower bracts, grow exceptionally well in the Florida landscape, as do lilies, vallotas, gloriosas, and many other flowering bulbs from Africa. And not to be overlooked are the ethereal flowers of the gingers, hovering in the garden like colorful birds in flight.

Flowering trees are also among the wonderful exotic plants that will thrive in Florida. Flowering en masse, to create a cloud of beauty against a periwinkle sky, Poinciana with its vivid orange canopy of blossoms is awe inspiring. Sphathodoceas, with their orange, tulip-shaped flowers, and Bauhinias, with their butterfly-like blooms, capture the eye.

All this beauty is readily available to Florida gardeners. With a little thoughtful planning and work, you too can create a tropical, exotic landscape unlike any other.

Designing the Exotics Garden

Designing the exotic garden is an adventure. Besides the usual rules of landscaping—unity, scale, and proportion—certain other conditions must be met. Because you are dealing with the unusual, you must use your imagination to capture the foreign beauty of your plants and create a vivid landscape. Proper plant selection and careful placement of plants and trees are vital.

Landscapes are usually either formal or informal; symmetrical or asymmetrical. The formal garden is symmetrical, with one plant balanced by another, and so on, often in simple geometrical patterns. A frequent example is a garden with orderly lines of clipped hedges enclosing balanced flower groups or trees. This arrangement can be beautiful but rather sterile. In the informal or asymmetrical garden, plants are scattered more or less randomly throughout the garden. Most exotic gardens are of an informal design. Forget about the spacious geometric gardens of France or Italy and look to more intimate Spanish and Moorish gardens for design inspiration. In either case, balance through the careful use of scale, proportion, and unity will achieve a lush and handsome tropical atmosphere.

Scale is the visual relationship of each form to every other form and to the overall design. Consider it a relationship of size. For example, a large house looks incongruous with a small entryway, and a cottage has a jarring appearance if an elaborate portico overshadows the dwelling. It is important to establish the correct relationship between the size of your home and the size of the tropical garden surrounding it. The starting point can be a tree, which will link the house and the garden. Avoid the common mistake of using too many palms, which tend to obscure the overall design. Small palms are certainly attractive in the average home landscape, but tall queen palms and other street palms will tower over your home and garden, thus disrupting the overall scale of the landscape.

Proportion is the harmonious relationship between each part of the total picture. A large paved terrace and a small lawn can be in proportion, or vice versa. But if both the terrace and the lawn are the same size, no visual interest catches the eye; one element fails to complement the other.

Unity results when all the pieces of the tropical garden and the home meld into an attractive-looking whole. You do not want a hodgepodge of unrelated masses insulting the eye. Plan on using related forms, colors, and textures to achieve unity in your garden.

Exotics, like all plants, have definite forms: spreading or horizontal, round or triangular, weeping or trailing. The form of a plant is important in the tropical garden; you should consider what the exotics will look like when mature, so that they will blend together yet be in proper scale and proportion to each other. Try to balance vertical forms with horizontal elements, round shapes with columnar. Try to paint a harmonious flow of plant material by using foliage, color, texture, and mass. But be careful about massing shrubs, which can grow quickly and crowd the garden; allow plenty of space between shrubs.

Other factors to consider when designing the exotic garden are rhythm—repeating the same group of plants or the same plant to give a sense of movement—and balance—placing plants of similar size, form, and emphasis in roughly equal proportions throughout the garden. Remember, scale, proportion, unity, rhythm, and balance each depend on the other; if you get the scale and proportion right, the other factors should fall into place.

Suit the design of your exotics garden to the site and style of your home. Follow a definite plan, and keep all parts of the landscape in scale. Create accent areas; select the exotics that will be appropriate for your particular conditions. Finally, but equally important, plan your exotic garden so it will need minimum maintenance. Use some paved areas perhaps, or easy-to-care-for ground covers.

Color

Exotic plants offer a large palette of color, but although the tropical or exotic landscape is lush, it is usually a green canvas with few colorful accents. Use colorful plants, such as *Strelitzia reginae* en masse or heliconias here and there for primary accents; complement these tall growers with low-growing plants such as cycads, aroids, or crotons (a favorite Florida landscape plant). Use bougainvillea to add drama to a fence, emphasize an arbor, or decorate a trellis. Bougainvillea can successfully frame a garden with color, and many varieties are available.

Select plants for your garden design with color in mind rather than specific plans or formal arrangements. The mood of your garden depends greatly on color combinations; red flowers next to yellow may be either jarring or invigorating. Blue flowers highlighted with white works to visually cool the garden, and for excitement use red, orange, and varying shades of pink scaling to white.

Plant Selection

Trees

Tropical trees are tempting additions of color and shade for the Florida garden, but because most properties in this state are small, too many trees will make the landscape crowded and unbalanced. Ask your nursery personnel about the height of any trees you are thinking of buying so you will be sure they suit the proportions of your landscape. (The tags on trees and shrubs usually indicate plant height at maturity.) A tree

Philodendrons are the focus of this tropical scene; plants climb trees but do not hurt them. Palms and various shrubs complete the picture, and in the background is a lath house for orchid growing. (*A. R. Addkison*)

Impatiens create a colorful flower bed in this Florida garden; bougainvillea covers the loggia at the right, and an areca palm is center stage. (*A. R. Addkison*)

Hibiscus, poinsettias (at Christmas time), and dusty miller create a colorful flower bed. Tall and stately *Hymenocallis* are in the background. (*A. R. Addkison*)

(*Barbara J. Coxe*)

such as erythina can grow as tall as 25 feet, whereas tabebuia grows only 10 or 15 feet high.

If you are planting a new garden, sketch the growing area and indicate the mass, height, and form of mature trees and shrubs; this plan will enable you to determine the proper balance for the garden. Place large trees at the front of the property, with smaller ones near the home. Remember that in hurricane country large trees can be a hazard, so plant cautiously: smaller trees are not as likely to be uprooted by high winds because they present less wind resistance.

Be sure any trees you plant do not interfere with foot traffic or obstruct paths.

When planting a backyard area, use a three-point design for trees: one small, two large, or two small, one large. When following this triangular design, place the large tree(s) at the rear of the property, with the smaller one(s) closer to the house. Use shrubs as fill-ins, and add some accent areas for further balance.

Florida gardeners can choose from a large selection of trees for year-round beauty. For spring flowers, choose bottlebrush, coral bean, jacaranda, and tabebuia. Summer-flowering trees include bauhinia, crape myrtle, golden shower, and royal poinciana. For autumn bloom try bauhinia or golden rain tree; and for winter color try African tulip tree, bauhinia, cordia, and Hong Kong orchid. (See Chapter 9 for more detailed information on exotic trees.)

Shrubs

Besides adding vivid color to the garden plan, shrubs unite landscape elements. Shrubs can be used as hedges, for fragrance, for foliage color and texture, or for the beauty of their flowers. Viburnum, podocarpus, pittisporum, orange jasmine, and small-leaved ficus add wonderful foliage accents, and gardenia, hibiscus, jatropha, medinella, and ixora are delightful when in flower. Select shrubs with the design elements of unity, scale, and proportion in mind, and carefully consider where you will place them in the garden.

When planting shrubs, less is more. Most shrubs grow quickly and can engulf a garden unless they are consistently trimmed and shaped to be kept in bounds. You do not want a jungle of overgrown plants. You want an orderly, tropical ambiance. Keep shrubs neatly trimmed, properly balanced, and judiciously shaped. These are favorites of mine:

Gardenia jasminoides bears a highly scented white bloom; a mature bush carries many flowers. The foliage is dark green, shiny, and very attractive. Gardenias love heat and humidity.

Hibiscus come in numerous hybrids, in orange, yellow, and other lovely colors. Size varies from small to large shrubs. These plants do well in the Florida heat and require little care. They are practically indestructible when cared for properly.

Jatropha pandiflora is a strong accent plant with brilliant flowers. It does not grow overly tall, and other than needing a somewhat shady growing place, it takes care of itself once it has been trained to keep in bounds.

Medinella magnifica is one of my favorites. In the Midwest I grew this plant as a container plant, but here in Florida it flourishes outdoors in a somewhat shady location. The pink flowers are a gardener's delight, but remember that this plant is extremely adverse to temperatures below 50 degrees F.

Ixora coccinea, or flame of the woods, is another favorite in Florida, and rightly deserves the many accolades it receives. This beautiful shrub can be used in many different locations in the landscape; it likes a somewhat shady spot. Chapter 10 lists many more exotic shrubs for the Florida landscape.

Flowering Bulbs

Bulbs native to South Africa and South America have long been overlooked by gardeners, yet they offer a large palette of color and need little effort on your part to thrive. Bulbs are wonderful accents in the landscape. Among the best flowering bulbs are amaryllis, eucharis, crinum, haemanthus, hymenocallis, vallota, lycoris, sprekelia, cannas, and alliums.

Bulbs flower in the summer or fall. They are hard to find at most nurseries, so buy from mail order sources (see the list at the back of the book). Once planted, most of these bulbs grow quickly if given adequate water and food.

When incorporating exotic bulbs into the landscape, one or two just will not do the job; you need masses of bulbs. Use flowering bulbs to create accent islands of solid color or between shrubs to carry the eye across the landscape. Do not use bulbs as added features. Instead, make them *the* feature and your garden will have the exotic ambiance for which Florida is so famous. See Chapter 11 for more details on growing bulbs.

Cacti and Succulents

It is not easy to landscape with cacti and succulents, but they do thrive in southern Florida. These plants come in the most varied colors and shapes imaginable, from tall and straight to candelabra-shaped and gnarled. Most cacti and succulents are desert plants, but the epiphytes of the family—the *Zygocactus*, *Selinocereus*, and others—are strictly rain forest natives that love the heat and humidity of Florida.

Avoid growing *Opuntias*, which are ungainly looking and quite invasive in the garden. Choose instead the statuesque *Trichocereus*, *Echinocereus*, and *Echinopsis*, whose shape, size, and growth habits fit in well with most small home gardens. Night-blooming cactus and the day blooming *Aporocactus*, with magnificent 7-inch flowers, are spectacular in the garden. These are overlooked gems that should be grown more often.

Succulents such as agaves and aloes flourish in Florida and with their wonderful sculptural shapes make terrific accent plants. Use them sparingly because they can grow as large as 5 feet in diameter. And do not overlook poinsettias; these beautiful shrublike plants—technically succulents—grow into a tree in southern Florida.

Most *Epiphyllum* and *Selenicereus* plants like to climb, so properly support the plant with trellises, posts, or fences. Don't worry if these plants start to crawl up a tree. They are not parasitic, and they will make the tree look good, especially when they flower in the summertime. Many hybrids and varieties are available through mail order.

Use caution when designing with cacti and succulents because most of these plants have spines and must be carefully handled. Getting a large specimen into the ground is a chore, so I recommend hiring nursery personnel to do the planting. These plants are superb singular accents near, but not close to, a driveway or door.

Cacti and succulents are covered in more detail in Chapter 18.

Bromeliads

Bromeliads have finally received their deserved recognition as landscape plants in Florida. I first wrote about these plants in 1967, and now, in the '90s, gardeners are fully aware of the many benefits of bromeliads. The plants are handsome landscape subjects and bear vibrantly colored flower scapes on tall spikes; all are easy to grow. Use bromeliads in groups rather than as single specimens.

Among the bromeliads, I recommend *Neoregelias*, *Aechmeas*,

Billbergias, *Guzmanias*, and *Vreiseas*. Beds of *Guzmanias* or *Quesnelias* will provide a panorama of summer color for little effort. The low-growing *Neoregelias* are wonderful ground covers, effectively replacing an expensive lawn. The swordlike inflorescence of the *Vriesea* adds outstanding drama in the garden. If you want impenetrable hedges, grow the viciously spiny *Bromelias* or *Hechtias*. Not even the bravest animal will be able to penetrate these barriers.

Bromeliads can be used as spot decoration on trees. To grow them on trees, anchor them to the trunk with moss and twine; eventually the bromeliads will take hold and provide unusual eye-level decorations. *Tillandsias* are frequently grown in this manner for a unique effect. (Bromeliads are not parasitic and will not harm trees on which they are grown.)

Bromeliads like sunlight and water; those are their only basic requirements. Some people think that bromeliads are breeding grounds for the mosquitoes that lurk in the water within plants' vase-shaped growth; this is an old wives' tale. However, frogs and salamanders may occasionally hang out in the vases, providing an interesting nature study for you and your guests. See Chapter 16 for more information on growing bromeliads.

Orchids

I have written five books about orchids in the past twenty years. I cannot say enough good things about orchids, which are fine landscape plants yet thrive equally well indoors as container plants in a garden room. The care and feeding of landscape orchids in discussed in detail in Chapter 17.

Vines

When I lived in northern California, I grew temperate-zone vines in my garden, and they were a constant source of delight. When I moved to southern Florida, I discovered the joys of growing tropical vines: Bougainvillea, allamanda, clerodendrum, pandorea, pyrostegia, and *Thunbergia grandiflora*. Tropical vines provide a welcome vertical accent in the garden, but as I learned from experience, designing with vines is not easy. These plants grow rampantly and are invasive, which means that you must constantly trim them to keep them in bounds. And if you plant vines against house walls, you will literally have to rip them away if you want to paint the walls.

I believe that the best way to use vines in the landscape is to grow them on arbors, pergolas, or trellises. This approach works because vines are crawlers and love to become entwined in the trellis-work and creep upward. You can buy trellises or lattices at home improvement stores, lumber yards, and nurseries or build your own from wood battens.

Once vines are in the ground, they are relatively easy to cultivate. Give plants ample water, some (but not too much) feeding, and judicious trimming. Chapter 12 details a number of wonderful exotic vines.

Gingers and Bananas

Only in the past few years, after discovering the terrific color these plants provide, have landscape designers begun using the many fine ginger and banana species available for garden use. Most gingers and bananas are easy to grow if they receive enough sunlight and water. Use the plants with discretion because they can be overpowering, looking garish rather than exotic if not displayed correctly.

The most famous members of the *Musa* family are the heliconias and hedychiums, both of which grace many Florida gardens. Used in mass, strelitzias and heliconias make a tremendously striking statement.

The ginger (*Zingiberaceae*) family includes tropical plants of stunning color and unusual design. Use costus, hedychiums, alpinias, curcumas, and zingibers as accent plants, three or four together in one area, to achieve a mass of color. *Alpinia purpurata* and variegated ginger have startling-looking flowers and handsome foliage. Give the plants some sun and ample water and they will grow with gusto. Some of the gingers can grow rampant and reach great heights, so consider these space requirements when planting gingers in the garden.

Most gingers and bananas are available mainly through mail order. See Chapter 13 for more information on growing gingers and bananas.

Water Lilies and Other Aquatics

Water lilies and lotus have been grown for centuries, and are experiencing a resurgence of popularity now that a renaissance in water gardening has begun. When in bloom, water lilies and lotus are indeed spectacular looking, but be aware that these plants require a good deal of care.

Other aquatic plants, such as *Cyperus*, are less demanding but just as attractive. These intriguing plants are especially appealing in the numerous watery locations prevalent in Florida, adding pretty interest to the overall garden plan.

The location of the water garden is most important: Be sure to establish the garden in an area that receives ample sun, affords you easy access, and is open to general view. You want to be able to sit back and enjoy this scene from a distance, so do not place the water garden too close to the residence. (Avoid the problem of mosquitoes by stocking the garden pond with koi fish and using the new, improved insect repellents.)

More about water gardens in Chapter 14.

Buying Plants

With the growing popularity of exotics, many Florida nurseries, garden centers, and even retail outlets now stock exotic plants, bulbs, and seeds. Mail-order houses specializing in tropical and semitropical plants offer the gardener an even wider selection of plants for the exotics garden.

Nurseries, Plant Shops, and Mail-Order Suppliers

Outdoor nurseries are the best places to purchase outdoor garden stock. Home and garden centers also offer garden plants.

Mail-order nurseries are fine places at which to become acquainted with the plants and learn about their cultural requirements. The owner is interested in selling the plants, and the variety is usually good.

If you can visit a nursery that has exotics by all means go and select your own plants. Look for healthy plants with stout stems and robust growth—nothing limp or wan. Avoid plants with yellow leaves or spots on them that might indicate disease. Inspect the growing medium; be sure there are no hidden insects and that the medium is fresh and not deteriorated with time. If the soil or fir bark is old it will smell sour or moldy. Shop for your plants as you would for produce at the market. Avoid any plant that has any sign of insect damage—nibbled leaf edges, holes in foliage, buds spotted, flowers off color—to avoid bringing home thrips, scale, mealy bugs, and other pests.

If you are buying bulbs, rhizomes, roots, or tubers check to see if they are firm and fresh looking rather than soft and discolored. Potted plants such as curcumas, hedychium, and kaempferias, should be in fresh dark soil and in clean containers. If the plants are not going directly into the garden, I generally repot all container-grown plants in fresh potting mix, because many times commercially grown plants are grown in soil-less media, which is cheaper, lighter, and more convenient for the grower. Repotting in new soil, whether in containers or the garden, is well advised.

In my home state of California there are shops devoted exclusively to bromeliads, orchids, and other exotic plants. In those I visited, most plants were in soil or on hanging rafts (pieces of bark). If you buy from a plant shop, ask the personnel what the plant is planted in, when it was potted, how old the plant is, and where it came from.

Also ask for plants by their botanical names; common names can be different in different regions. Generally, plant shops are fine places at which to buy indoor or container-grown exotics because most of the owners are knowledgeable about their plants. Pots should be tagged with identifying markers; if they are not, ask for the botanical name.

If you cannot get to a nursery, do not hesitate to order what you want from a mail order catalog. With improved shipping materials and methods there is little risk of losing plants in transit.

Most plants from mail-order suppliers are shipped bare-root or in pots. If possible, buy plants in pots. This means extra weight and thus more shipping cost, but because bare-root plants have been uprooted from their pots and then sealed in closed boxes for travel, by the time you get them they can be in sad shape. Most plants recover in time, becoming fine specimens.

When ordering from mail-order companies, most exotics will be sent as bulbs or rhizomes, sometimes as potted plants. When you get your shipment unpack it as quickly as possible to prevent spoilage. Plants should go into the garden or into potting soil immediately. Set bulbs or rhizomes on a table for a day or so exposed to air and then plant them immediately. Most plants, bulbs, or rhizomes are shipped at the proper planting time for your region. Seeds can be ordered at any time. A list of mail-order nurseries appears in the appendices.

Cost

With the exception of new introductions or rare species or varieties, exotics are generally inexpensive. Be prepared to pay about $10 for a mature exotic and $5 for a seedling. If you are a beginner, start with mature plants; later, after you have some growing experience, buy

This screened enclosure depends on an *Howea forsteriana* to accent the pool; angel wing begonias and orchids furnish needed color. (*A. R. Addkison*)

A poolside setting with vanda orchids in wood baskets. Tall dracaena and meandering palms create a tropical picture. (*A. R. Addkison*)

Two large dracaenas set the stage for this patio, and the focus is a Phaius orchid in a clay container on the table. (*A. R. Addkison*)

seedlings. If prices are higher than $10 for popular varieties, you are buying from the wrong source. Shop around; look at more catalogs.

Be especially wary of advertisements for collections of several different kinds of exotics for $10 because invariably these are inferior plants a grower wants to get rid of. And be especially cautious of full-page ads promoting a specific plant at a low price: It simply does not make sense to offer a "bargain" yet pay thousands of dollars for a full-page ad.

Arrival At Home

When you get your plants home, check them again to see if any insects are present. If plants are bare-root, inspect the roots for insects. Also check mail-order plants upon arrival because insects can travel great distances and still survive. Sometimes plants carry invisible insect eggs rather than mature insects. In any case, soak each plant up to the pot's rim in a sink full of water for an hour, causing any uninvited guests to come to the surface. Then spray plants with a strong jet of water to dislodge any eggs. Finally, polish leaves with a damp cloth. Do not use any leaf-shining preparations because all they do is close the pores of the leaves.

Do not put new arrivals in direct sun because the abrupt change in light can harm them. Place the plants in a shady location and water them. In a few days, move the plants to a somewhat brighter place for another five days or so. Finally, move plants to their permanent sunny or bright location. During this time, be sure the plants have good air circulation.

Two

GROWING PLANTS

Climate

Florida can be heavenly during the winter months but sweltering during the summer. The state boasts the highest average winter temperatures of all the states and the sunniest winter climate in the eastern part of the United States. Southern Florida is tropical below Melbourne on the east coast to Fort Myers on the Gulf coast, which means that the average monthly temperature is about 65 degrees F. There are distinct wet and dry seasons. In the wet season, temperatures are warm, humidity is high, and rainfall occurs just about every afternoon (up to 15 inches or more of rain in the summer). In the dry season, sun is abundant, temperatures and humidity are lower, and little rain falls. To survive here, plants must be able to tolerate both heavy summer rains and partial drought.

North of the Fort Myers-Melbourne belt, Florida is considered subtropical or temperate. This area does not generally have sharply defined dry and wet seasons, and monthly temperatures can fall well below the 65 degree F limit of the southernmost part of the state.

Plant Hardiness

The United States Department of Agriculture categorizes plants according to cold hardiness; hardiness zone maps produced by the

USDA are used by nursery personnel and gardeners to determine what will grow where. The maps are based on average minimal night temperatures at weather stations. These temperature or hardiness zones are helpful for predicting the adaptability of plants to specific climates. Florida is divided into zones 8, 9, 10, and 11. The zone maps are helpful, but they do not and cannot show all the subtle differences in the microclimates that occur along bodies of water, the amount of wind or rainfall, the length of days of sunshine, and even the flooding and freezing that occur in Florida.

A hardy plant survives the climactic extremes of a particular area, even freezing, by binding its water against low temperatures. Plants that freeze form ice within their tissues. If the temperature drops quickly, plant tissue freezes quickly, ice crystals form, and the plant dies. If tem-

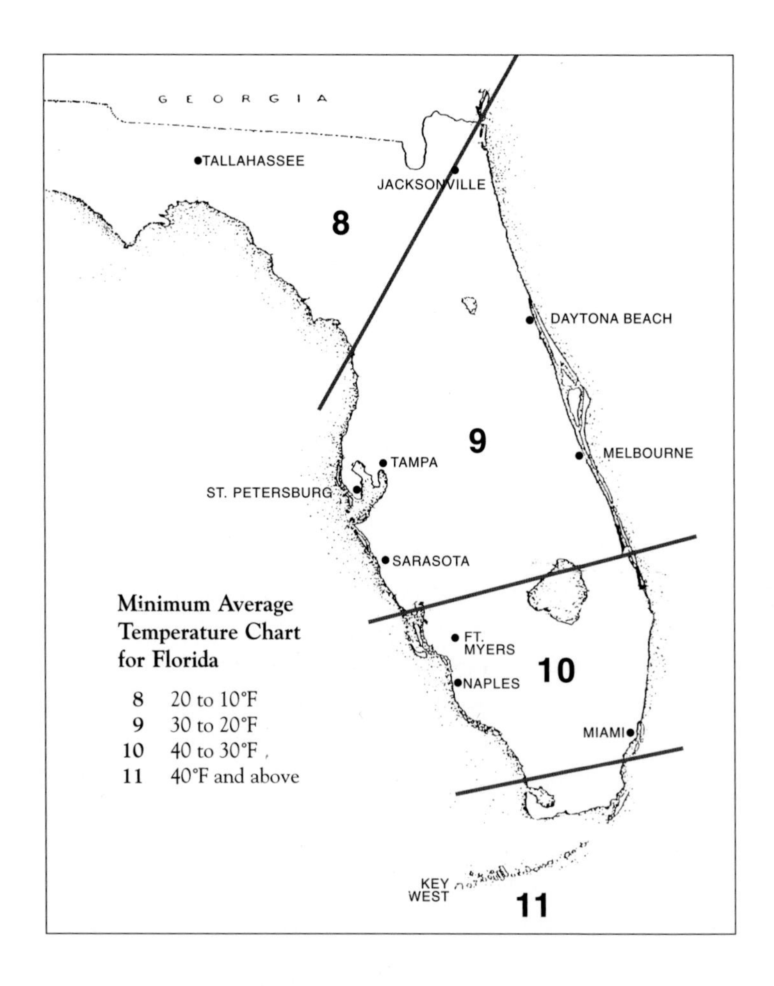

peratures fall gradually, plant tissues freeze slowly but may recuperate if freezing does not occur too often.

Tropical plants are classified as those that will thrive where temperatures do not drop below 45 degrees F. If they are grown at temperatures below 45 degrees, they must be covered with burlap or other materials (see Chapter 3) to protect them from the cold. For plants to be protected from freezing, they must make steady growth throughout the growing season, with a gradual cessation of growth from mid-August until the first killing frost. Fertilization and watering must be carefully administered after midsummer. Plants that lose their leaves due to insect infestation or disease are very vulnerable to premature freezing.

Late feeding with high-nitrogen fertilizers can force late succulent growth and reduce the hardiness of plants such as hibiscus, bougainvilleas, and other exotics. Generally, it is best not to feed plants in the late summer. Permanent mulching can also reduce a plant's hardiness by keeping the soil temperature warm late in the season. Remove mulches in the fall and replace them before actual winter weather begins.

Soil moisture is also important for plant hardiness. Evergreens and deciduous trees and shrubs lose water through their stems all winter. Winds speed up this moisture loss. The stronger the wind, the greater the loss of water. When the ground is frozen, plants cannot absorb water and thus replenish the lost moisture. Mulches can help in this situation.

Soil

Topsoil is composed of small particles of disintegrated rock and minerals; decomposing organic matter; living organisms such as bacteria and fungi; water, which holds the dissolved mineral salts; and air. *Subsoil*, which is beneath the topsoil, contributes to fertility and determines the drainage characteristics of the soil. Soils are sandy (light and quick draining), loamy (porous, so that water drains moderately), or claylike (heavy and slow draining). A good, fertile soil is a mixture of clay, sand, and humus. Over the years, most cultivated soils lose their mineral content, so they must be reworked and enriched with the addition of organic matter or *humus*.

To make most Florida soils fertile—porous in texture, with good drainage, and spongy to retain moisture—humus must be added. Humus is decayed organic matter. Animal manure, compost, leafmold, and peatmoss are all forms of humus. Humus adds body to topsoil and provides aeration; as it dissolves in the soil, it provides nourishment for plants and microorganisms. Humus is used and depleted constantly and must be replaced. How much depends on the soil, the kind of plants grown, and the existing content of humus. For most gardens, the addition of about 1 inch of compost to the soil annually is sufficient.

Florida is not blessed with the best soil. The eastern or Gold Coast's soil up to about Lake Okeechobee is an alkaline (calcareous) rockland that supports mainly sawgrass and mangroves; the soil here contains large amounts of sand and limestone. The southwestern part of Florida, up to near Tampa, has sandy soil with few nutrients. The middle part of the state has areas of dark and sandy subsoil and poor drainage. This is the area where most of the citrus is grown. North of Tampa, to Ocala and Gainesville, the soil is a combination of heavy sand and loam with some limestone and good to fair drainage. The panhandle area has dark, sandy subsoil. The extreme northern part of the state has good to fair soils for growing; soil here is generally deep and well drained.

Fertilizer

Not only must humus be replenished in soil, plants also need applications of fertilizer, containing nitrogen, phosphorus, and potassium (potash). Trace elements such as copper, iron, manganese, sulfur, and zinc are also important to plants. *Nitrogen* stimulates vegetative development and is necessary for the growth of stems and leaves. *Phosphorus* is needed in all phases of plant growth, and it also produces good root development. Foods with a high content of phosphorus, such as 12–12–12 or 5–10–5 make plants bloom. *Potassium* promotes a plant's general vigor, making it resistant to certain diseases. Potassium also tends to balance other plant nutrients.

Feed most plants in the spring, when they begin growth, and in the summer, when they are actively growing. Never feed new plants: Wait a few weeks until they overcome the shock of transplanting. And never feed sick plants because they are not strong enough to absorb more food. For all plants, frequent weak feedings are better than one massive dose that can burn roots and foliage.

To feed shrubs, vines, and flowers, spread dry or granular fertilizer on the ground around the plant and lightly scratch the food into the soil. Water thoroughly to dissolve the fertilizer; unless dissolved in water, roots will be unable to absorb the nutrients. Feed trees through holes punched in the ground. Use *foliar feeding* for small plants; this is a fertilizer solution applied to the leaves—it is instantly available to plants but not long-lasting.

The ratio of nitrogen (N), phosphorus (P), and potassium (K) in a commercial fertilizer is denoted on the package or bottle in that order. For example, a bag of 10–10–5 fertilizer contains 10 parts nitrogen, 10 parts phosphorus, and 5 parts potash. There is also a portion of inert filler.

Fertilizers are classified as organic (made of natural substances) or inorganic (made of chemicals). Organic foods such as blood meal or bone meal improve the structure of soil and release nutrients slowly. Organic fertilizers slowly release nutrients over relatively long periods, using soil bacteria to decompose and transform the resulting compounds into the nitrogen plants can use. Organic fertilizers include sewage sludge, animal and vegetable tankage, manures, and other natural materials.

Inorganic compounds are available in five forms: *powdered*, which tends to blow away, stick to foliage, or cake if stored in a damp place; *concentrated liquids*, which are good all-purpose fertilizers; *concentrated powders*, which are diluted in water and applied to foliage and roots; *concentrated tablets*, which are recommended mainly for houseplants; and *pellets* or granular fertilizers, which are easy to spread. (Some granular fertilizers also contain insecticides or weed killers.)

Ureaform compounds are synthetic materials made by the chemical union of urea and formaldehyde. Ureaforms slowly release nitrogen; urea (an organic fertilizer) quickly releases it, so do not confuse the two materials.

Some fertilizers are made for specific plants, such as roses or camellias. Fertilizers produce better growth and flowers when they are used wisely; used with abandon, they can kill plants.

Soil pH

In addition to good drainage and fertility, your soil must have the proper pH, which is the measure of acidity or alkalinity of the soil. You can purchase pH test kits at most garden centers or have your soil tested by a lab. Neutral soil has a pH of 7; below 7 the soil is acid, and above 7 it is alkaline. Most plants thrive in neutral or near-neutral soils. In alkaline soil, potassium becomes less and less effective and eventually becomes locked in. In very acid soil, aluminum becomes so active that it becomes toxic to plants. Acidity governs the availability of nutrients in the soil, determines which bacteria thrive, and to some extent affects the rate at which the roots can take up moisture and leaves can manufacture food.

To lower the pH (increase the acidity), apply one pound of ground sulfur per 100 square feet. Spread the sulfur on top of the soil and then water. To raise the pH (sweeten the soil), add 10 pounds of ground limestone per 150 square feet. Scatter the limestone on the soil, mix it in well with the top few inches of soil, and water it in. Add the limestone in several applications at six- or eight-week intervals rather than all at once.

Mulches

Mulching decreases the amount of moisture lost through evaporation from the soil surfaces, keeps the soil cooler than when it is fully exposed to the sun, and helps control weed development. Any material spread between and around a plant to cover the soil is a mulch. Inorganic mulches such as stones, gravel, plastic, and paper do not help improve the soil. Most mulches are organic materials that decay in time and contribute to the improvement of the soil.

Old leaves are superb mulches; oak leaves and pine needles are perfect for acid-loving plants; straw decomposes slowly and is free of weeds; and grass clippings are good when mixed with a coarse material so that they do not mat. Cocoa beans and pecan shells decompose slowly and are fine for all but acid-loving plants. Ground fir bark comes in several grades and is attractive; it stays in place and decomposes slowly. Sawdust is fine used alone or mixed with peatmoss.

Apply loose mulches in a 2- to 4-inch-thick layer. You can keep plants mulched most of the year, or apply the mulch in the spring after the soil has warmed up and growth has started. If mulches are put in place too early, they stop growth by keeping the soil cool.

Water

Soil absorbs water more slowly than you think. After 1 hour of normal watering, sandy loam soil will be penetrated only to a depth of about 30 inches, and clay with loam to a depth of 12 inches. Thus, 5 or 10 minutes of watering will not help your plants much. Water must penetrate the soil deeply to get to the roots below the surface. If only the top of the soil is kept wet, roots become shallow because they cannot work their way into the deep soil to look for more moisture. Shallow roots are susceptible to heat damage.

WATER PENETRATION RATE BY SOIL TYPE

Depth	Sand	Sandy loam	Clay with loam
12 inches	15 minutes	30 minutes	60 minutes
30 inches	40 minutes	60 minutes	
48 inches	60 minutes		

As important as watering is, it is equally important to let plants dry out before the next watering. Too much water in the soil blocks oxygen to the roots, and plants begin to drown. In other words, the important part of watering plants is not how much but how much *when*. The *when* depends on the wind, temperature, intensity of light, humidity, soil, and rainfall.

Clay soil holds water longer than sandy soil; sandy soil drains very quickly; loamy soil falls somewhere between clay and sandy soil. For example, to thoroughly soak a 50-square-foot area to a depth of 24 inches, sandy soil takes 60 gallons, loamy soil needs about 100 gallons, and heavy or claylike soil takes almost 175 gallons. A hose under normal pressure runs about 5 gallons per minute; it takes about 15 minutes to soak sandy soil, 20 minutes for loamy soil, and 40 minutes for clay soil. The sandy soil will dry out in 7 to 9 days, the loamy soil in about 14 days, and the clay soil in about 20 days.

You can water plants with sprinklers and hoses, bubblers, and various drip irrigation systems. Sprinklers and hoses are fine in the average garden. When using a sprinkler, it is important to know how much water it actually delivers. To determine this amount, place an empty coffee can on the ground and note how long it takes the sprinkler to fill the can with 1 inch of water. Multiply this by the number of inches you want the water to penetrate; leave the sprinkler on that amount of time. Be sure to really soak the soil when you water plants and trees.

Sprinkling is the best way to apply water evenly over a large area, and most tropical and subtropical plants thrive with overhead watering because dust and soot are washed off the leaves. The moisture also discourages certain insects from attacking foliage.

Drip system watering is another way of getting moisture to your plants, and if you have an extensive garden this is a fine method. Drip system kits are sold at nurseries and hardware stores and can be easily installed either professionally or by yourself. Drip system watering applies specific amounts of water to specific areas around the plants.

Three

Protecting Plants

Southern Florida (Sarasota to Miami) rarely gets freezing weather, but the north and central areas (Jacksonville, Gainesville, and Tampa, for example) occasionally do go through a cold snap. The below-freezing air from the North pushes down and squeezes out the warm air, chilling plants dangerously. Wind storms and heavy rain may also damage plants, but there are ways you can combat Nature's temperament and bring plants through hard times.

Freezing and Thawing

When planning your garden, consider that plants growing in a south exposure will be more protected from cold weather than those growing in other locations.

Freezing and thawing adversely affect most, but not all, exotics. Many philodendrons, for example, can survive a period of bad weather. You have to wait and see what happens. Do not immediately start cutting down trees and hacking off branches; like other plants, the exotics have a strong will to survive, and many will come through adverse periods. The recovery may take weeks or even a month before full vigor is restored.

During the winter, listen to weather reports that alert you to any upcoming cold snaps and prepare to protect your plants. To protect small or medium-sized plants, sink twigs or sticks into the ground around the plants, in a teepee style. Put a sheet or bedspread over the teepee frame, and then weight down the covering with stones or bricks to prevent it from blowing off.

Even easier than a teepee or tent is to drape plants with plastic tarps (sold at hardware stores). With cord, tie the tarp around the base of the plants. Be sure to punch a few holes in the tarp so the air can circulate around the plants to prevent scalding or wilting as temperatures begin to rise again.

Burlap is excellent for protecting plants from freezing temperatures. (You can buy burlap at surplus stores or through mail order garden suppliers.) Burlap is sold in sheets or rolls; cover the plants and tie the burlap around the base of the plants. Large cardboard boxes, punctured for air circulation and inverted over plants, also protect them from freezes.

During periods of light to moderate frost, a gentle continuous spray of water over plants will effectively protect them, but in heavy frost the water may solidify into ice and damage the plants.

After a period of freezing weather has passed, do not cut away dead leaves or branches. Wait until new growth appears before doing any pruning. Do not fertilize plants after a freeze until they are actively growing again, which may not occur for several weeks.

Large, established trees and shrubs usually can cope with freezing and thawing periods without protection. Despite damage they eventually start new growth.

Heavy Winds

Winds can play havoc with trees. Avoid planting trees with shallow roots or brittle limbs, like *Albizzias*, *Grevilleas*, some *Acacias*, and *Spathodea*. Ask nursery personnel about the sturdiness of any trees you are considering buying. (Most nurseries that sell large trees have qualified people to help advise you.)

Not all trees that are knocked down by heavy winds are gone forever. Uprooted trees can survive many weeks and be righted if their roots are protected. Cover the roots of a downed tree with old blankets; keep the blankets damp so the sun does not burn the roots.

If branches are broken or down, clean up the mess and let the tree recover—it will. Trees that have more than 60 percent of their canopies gone or branches split or cracked most likely will have to be taken down and removed.

After a wind storm, assess the damage. Save as much of the green on the tree as possible. Prune judiciously; worry about fine pruning and shaping later. If a tree has been set askew by the wind, stake it back to its original position. This work is not as hard as you may think. First, pound three stakes (tree poles sold at garden suppliers) into the ground around the tree. Space the stakes equally, creating a triangle around the tree. Tie a piece of rope to one of the stakes, and then run the rope around the tree. Place sections of an old hose around the guy rope where it strikes the tree bark, to cushion and protect the bark.

After staking a tree into its original position, water gradually. You do not want to try to force the tree to grow. Just be sure the roots obtain enough moisture. If you insist on feeding the tree after a wind storm—I do not recommend this—use small amounts of a plant food heavy in potassium, light in nitrogen.

After a wind storm, shrubs may be somewhat damaged. Usually, only reshaping is needed, so lightly prune the plants. Exotic plants such as the gingers, bananas, and some *aroids* may be ripped apart by wind and die down, but new growth will start when the weather is settled. Actually, low-growing exotics tend to grow more luxuriantly after a wind storm because trees and branches that obscured the sunlight reaching them have been removed.

Prevention

Always keep tree branches thinned out as they become ungainly. Topping trees and keeping them growing upright so air and light can pass through the network of branches will go a long way in preventing downed branches and limbs when storms hit. However, do not top and prune large trees yourself. It is much safer, and the results will be more accurate, if you hire professional tree trimmers.

Keep shrubs clipped and shapely. You do not have to worry about perennials such as the heliconias and the gingers because they will recover even if they are ripped to the ground. Orchids and bromeliads deserve special attention because most of these exotics just cannot take freezing or wind. Take containers of orchids and bromeliads indoors. For those exotics growing in the ground, protect them with a cover of plastic tarp, burlap, or a sheet. If some orchids do die back after a period of frost, wait and see if they recover before removing them.

In 1989, southern Florida was hit with a bad freeze. My garden looked bleak, but within a year, almost every plant had recovered on its own. Today these plants are all thriving. Remember, keep plants healthy during good weather; after bad weather, wait and see. Assess the damage and take the proper precautions. Give your plants a chance to regrow on their own, and give them loving care to help them recover, and in most cases they will.

Four

INSECT CONTROL

No matter where you garden, you will have to deal with insects. Many years ago, when I lived in northern California, I had to battle just as many bugs as I do here in Florida. But there are safe and effective ways you can keep the pests under control—proper care, good cultivation, and cleanliness in the garden will ensure a pest-free environment unharmed by the toxic effects of chemicals.

Insects live within a well-structured society, and they are vital to the balance of nature. Some insects, such as ants and beetles, are beneficial to the soil because as they burrow through the soil they leave openings through which air can penetrate. The air helps improve the structure of the soil as well as its drainage. Other insects kill weeds, and many insects are pollinators. It is indeed true that much plant life on earth would disappear were it not for insects.

On the other hand, some insects will injure rather than help your plants, and you will have to take action to eliminate these destructive forces. If you are able to recognize the dangerous bugs (we describe them at the end of this chapter), you can control the enemy naturally with the help of beneficial predators and parasites rather than poisons.

Natural Controls

Natural control involves using the predators, parasites, and insect diseases normally present to curb the population of the destructive pests.

(Biological control is using insects that have been raised in laboratories or insectaries.) You thus work with Nature, not against her.

The *ladybug* is a beetle that feeds on aphids, mealybugs, whiteflies, scale, and the eggs of other insects. Ladybugs are red to reddish yellow to tan or brown with black spots; some are black with red spots. Their bodies are deeply segmented. These predators get into places impossible to reach with a spray of water or chemical; the average adult can eat up to almost 400 aphids a day.

If your garden has been kept chemical-free, larvae of the ladybugs will be present in the early summer. Protect the larvae (which do not resemble the adults at all) because they are such superior predators. Ladybug larvae are dark-colored and clumsy looking, like an awkward caterpillar.

You can also buy ladybugs from suppliers. The insects are housed in small cases. To place the bugs in the garden, dampen the ground and set them in place by the handful, near a food supply (aphids).

Aphis lions, widely distributed throughout the United States, include the ant lions or doodlebugs, dobson flies, and lacewings. Aphis lions are some of the most helpful garden predators; 15 families make up the species. The insect captures, punctures, and sucks juices from aphids, mealybugs, whiteflies, scale, moth eggs, caterpillars, and sometimes, when it is very hungry, thrips and mites.

Lacewings are soft-bodied, with veined and gauzelike wings, either green or brown. They like red spiders in addition to the foods mentioned above. They are usually nocturnal.

Doodlebugs trap mainly ants by digging a pit and burying themselves at the bottom of the pit, where they wait for victims to fall in. The adults look like damselflies (damselflies prey on small, soft-bodied insects).

Dragonflies are brilliantly colored and have enormous eyes that can see in all directions at once. They are extremely fast and maneuverable fliers. The green dragonfly is the most common. Dragonflies eat all kinds of insects.

Ambush bugs dwell in flowers; they hide behind the foliage or blooms and grab their victims. The bugs are very ugly, but do not kill them because they are beneficial. The *assassin bugs* are equally ugly, and their bite is as painful as a wasp sting. Note that both ambush and assassin bugs may also eat some good insects, so keep an eye on them.

Damsel bugs are excellent predators, eating aphids, mites, caterpillars, and other pests. *Pirate bugs* are hard to see because they are so small, but they control aphids, scale, thrips, and other garden pests.

The *praying mantis* can be bought from suppliers. When young, these insects live on soft-bodied insects such as aphids and leafhoppers. When they are mature, they add tent caterpillars, chinch bugs, and beetles to their diet. Mantids kill other insects by using their sharp beaks, mandibles, or jaws. Their front legs hold the prey. Buy praying mantis cases between November and May. Tie or tape one case to a shrub or tree, at least 2 to 4 feet aboveground. The insects will emerge sometime in June. You need about five egg cases per half-acre. Mantids will stay in the garden as long as their food supply lasts.

Spiders, of the class *Arachnida*, are not true insects, but are beneficial predators in the garden because their main diet consists of live insects.

Predators eat their victims; parasites enter the body of the host insect and damage the host's tissues. When the insect dies, the parasite may continue living in the dead body, or it may emerge and pupate elsewhere.

The *parasitic wasps* of the order *Hymenoptera* are the most valuable, destroying the larva of moths, butterflies, beetles, and aphids. The ichneumon fly, really another wasp, is a tremendous killer of the gypsy moth.

Chalcids consist of thousands of species. These parasites are tiny, black or yellow, with a metallic sheen. They attack egg, larval, and pupal forms of scale, mealybugs, aphids, flies, and beetles. Some chalcids feed outside the host insect.

The *trichogramma* is the most important wasp in the garden; it is available from insectaries. This parasite attacks more than 200 different kinds of insects. When trichogrammas arrive from the supplier, they are almost ready to hatch. Place opened containers in the areas you want controlled (one container has 2,000 to 4,000 parasites, enough for 5 acres). Use these insects wisely because they also attack butterflies.

Birds

Birds are superb insect eaters; insects comprise many birds' main diet. For example, a chickadee eats 200 to 500 insects a day, a brown thrasher can eat more than 600 insects in a day, and a house wren feeds 500 spiders and caterpillars to its young within a day.

To attract birds to the garden, provide food, water, and protection. Packaged food is sold at supermarkets and pet stores, and bird feeders come in a variety of attractive styles. (Do not feet birds suet, which is a substitute for larvae; you want the birds to eat insect larva.) Bird houses are nice protective places, but because some birds are quite particular about the type of accommodations they prefer, choose the house carefully.

Trees and shrubs give birds a place for nesting, protection from natural enemies, and a source of food (insects), so work these plants into the landscape. And birds love water, for both drinking and bathing. Have many forms of bird baths in the garden because different birds like different depths. (Bluebirds love just a mist of water.)

Be sure the bird baths are not too steep-sided and not more than 3 inches deep. Sides should slope down gradually, so the birds do not fall off. The surface of the bath should be rough, to provide good footing. If you have cats as pets, or to prevent harm from neighbors' or feral cats, remember to keep the bath high enough so cats cannot get to it. The bath should also be in the open so birds can see any enemies.

Among the helpful birds in the garden are chickadees, house wrens, phoebes, bluebirds, robins, cuckoos, and woodpeckers. Swallows depend almost entirely on insects for their sustenance. The purple martin is the most useful swallow. Baltimore orioles eat caterpillars, beetles, ants, grasshoppers, and click beetles; cuckoos like hairy caterpillars, beetles, grasshoppers, sawflies, stinkbugs, spiders, tent caterpillars, and crickets. The kingbird eats almost all kinds of harmful insects. Woodpeckers prefer wood-boring beetles and all fruit- and fruitwood insects. Towhees pick off hibernating beetles and larva. Meadowlarks also eat some weeds in the lawn. House wrens, chickadees, and titmice love beetles, caterpillars, flies, and plant lice and scale.

Harmful Insects

Observation is the watchword for a healthy, insect-free garden. Be on constant lookout for bad insects, such as aphids, leaf hoppers, beetles, and other destructive critters. Natural controls and the use of birds may be all the "ammo" you need to keep your garden safe.

However, sometimes you may have to resort to other means if natural controls do not work, as I do to get rid of the fire ants that occasionally invade my garden. But remember, if the natural controls of using predators and parasites do not work, consider the nontoxic methods of eliminating pests, such as a solution of water and laundry detergent, before resorting to toxic chemicals. Remember that the use of chemicals may kill beneficial insects too. Here is a quick rundown of common garden pests and safe and natural ways to get rid of them.

Aphids

Sometimes called *plant lice*, aphids pierce stems, leaves, buds—practically all parts of a plant—and extract vital plant juices. The leaves curl and growth becomes stunted or stops altogether. Aphids give off a sort of honeydew that attracts ants, which love this type of food. The ants move the aphids around the garden; where there are ants, there soon will be aphids. If plants are attacked by a very heavy infestation of aphids, the plants usually die. Aphids also sometimes introduce viruses into the garden.

Most aphids are black or green, sometimes red. They breed incredibly fast and in large numbers, reproducing continuously in Florida's warm climates. Ladybugs and praying mantids are superb natural predators of aphids. For heavy infestations, apply a solution of half water and half laundry detergent to get rid of the aphids.

Leafhoppers

These small wedge-shaped insects suck sap from leaves, making leaves pale or brown and stunting plants. Some species contribute to the spread of viral diseases. When disturbed, leafhoppers hop or fly away. They excrete a sweet honeydew that attracts ants. Use a good botanical spray, such as Pyrethrum or Rotenone, to control these insects.

Thrips

These slender, needlelike, winged insects scrape stems or leaves and then suck out plant sap. Leaves become silvery. The nymphs are wingless. Thrips hibernate during the winter and begin feeding on young plants with the advent of warm weather. They can produce a new generation every two weeks in hot weather, so do not let them become established. Use insect predators such as toads for control, or spray plants with a blast of water to dislodge and kill thrips.

Scale

Scale may cause extensively damage to shrubs and flowers. These tiny oval-shaped insects are either soft-shelled or have an armored hull. Scale are hard to get rid of because they stick stubbornly to leaves and stems. They suck sap from plants and can kill a healthy plant very quickly. Like aphids and thrips, they produce a honeydew that attracts ants. Use dormant oil sprays to control scale.

Mealybugs

These tiny, white, cottony insects hide in leaf axils and form deadly colonies that can destroy plants. The males can fly; the females are wingless. Dip cotton swabs in alcohol and dab the insects if you have a mild infestation; use insect predators for severe invasions.

Lace Bugs

These small insects with large, lacy wings suck juices from plant leaves and stems and make leaves look mottled; they leave dark, shiny excrement on plants. Lace bugs overwinter as eggs attached to leaves and in warm weather produce their broods. Control them with botanical sprays, such as Pyrethrum or Rotenone.

Leaf Miners

Leaf miner larva feed and live for part or all of their lives on a leaf. The miner feeds on the surface of the foliage, defoliating a plant. Miners imprint long and narrow or blotchy designs on leaves. The best protection is to kill the adults before they begin laying or thwart the larva as soon as they begin feeding with a laundry soap and water solution. Keep the area around susceptible plants free of debris where adults can hide. Leaf miner adults can be sawflies, moths, or flies, many of which can be controlled with light traps. Be aware that light traps also kill beneficial insects along with the bad ones.

Borers

Borers dig even deeper than leaf miners; they are the larval stage of certain beetles and moths. They attack the hard woody parts of a tree or shrub, the softer tissue beneath the bark, or even soft-stemmed herbaceous plants. They cause wilting and holes in stems and branches. Borers are very hard to see, and sun helps aggravate the insect damage and hastens the wilting of a plant. You must attack borers at the proper moment: Kill the newly hatched larva *before* they have a chance to enter plants; otherwise, it is almost impossible to kill them.

Insect Traps

Insect traps are quite useful in the garden, but they do capture good insects along with the bad ones. Boards placed on the ground will attract squash bugs, snails, and wireworms, which will crawl under the boards to hide. Lift up the boards and kill these pests. Other effective traps are trenches and ditches, which will stop migrations of insects; pickle jars filled with molasses and water; tubs of kerosene; and pine tar and molasses on a cardboard band. All these devices will lure insects to their death.

Light Traps

An ordinary light bulb attracts night-flying insects. This principle is used in commercially made light traps; most have a small motor and fan that draws the bugs into a removable bag and some light traps include a voltage grid that electrocutes the insects. Red and yellow light does not seem to attract many kinds of insects, so clear blue, violet, green, and even black lights are often used. Other factors that affect the effectiveness of light traps are the height of the trap and the climatic factors of temperature, humidity, and wind direction.

Many traps are attractive enough to hang near the house or garden. Some traps are portable; others have to be mounted permanently. Automatic timers or electric eyes turn the lights on at dusk, off at dawn. Remember, however, that traps kill good bugs along with the bad ones.

Vacuum net traps that operate on the same principle as a household vacuum cleaner are also used.

Five

LANAIS, LATH HOUSES, AND SWIMMING POOL AREAS

Lanais and lath houses are familiar structures in tropical landscapes and these partially sheltered areas are excellent for housing orchids, gingers, and other exotic plants. The lanai is usually an open construction with or without a partial roofing and the lath house is a ventilated area protected from strong sunlight by wood lathing.

Screened enclosures are very popular in southern Florida to enclose pools or patios. Screened structures in themselves can be a marvel of design with butterfly roof configurations or geometric designs worked into the shape of the enclosure. Exotic plants can make these outdoor areas inviting and dramatic.

Whether you have a screened enclosure, lath house, lanai or swimming pool area, pay attention to sun angles. If the open end of an enclosure faces west, provide some protection from the direct rays of the sun by using roll-up blinds or other screening materials. Trellises or awnings work well to give plants solace from the hot sun, especially in summer. Eastern exposures are best for most plants; here they can receive sunlight without fear of leaf scorch; southern sites are generally all right for most light-loving plants. Don't neglect northern exposures—many shade-loving tropicals grow well in these areas, especially certain orchids and bromeliads.

A reflecting pool with tall elegant Cyperus makes a handsome sight; the patio paving is handsome, an ideal surface for the lanai. A Livistona palm is at the entrance to the house (in background). (*J. Kramer*)

Lath houses are a frequent sight in southern Florida, where orchids and various tropical exotics thrive. This Oriental design adds dimension to the garden. (*J. Kramer*)

Strelitzia nicolai is a Florida landscape plant that is frequently used in lanais, gardens, and screened enclosures because it is easy to grow and affords a lush look to the design. (*J. Kramer*)

Shell ginger is the center attraction in this Florida garden. (*J. Kramer*)

The Lanai

In Florida the lanai can be many things: a traditional porch, an open-sided roofed area for entertaining, or a semi-enclosed room. Lanais are sometimes called Florida rooms, so closely are they associated with the tropics, although the word itself is Hawaiian.

The lanai, generally adjacent to the living room or dining room, acts as an extension of the living area. This area should receive a careful selection of container-grown plants like hibiscus, gardenias, bougainvillea, and gingers. Use large plants such as scheffleras and plumeria, alpinias and anthuriums (dozens of varieties) to create background "walls" of attractive foliage.

Lath House

I am particularly fond of lath houses because they offer an almost perfect environment for exotic plants such as orchids and gingers. Here, with alternating patterns of sunlight and shade, plants receive just the right amount of light and you never have to worry about too much or too little sun exposure. A lath house is a partially enclosed area made of 1 x 1 or 2 x 2 lathing (sold in bundles at lumber stores) spaced an inch or two apart to form open walls and a roof. The lath spacing filters the sun throughout the day. Floors are generally made of cinders or gravel. The lath house should be large enough for benches so that people can sit and take time to smell the flowers and enjoy the shade.

Screened Enclosures

These large, intricately designed roofed structures offer a great stage for the dramatic use of exotics. Because there is partial protection from the elements in the screened area excessive protection is not needed unless temperatures drop below 35 degrees F for several successive nights. Large-leaved gingers, bananas, and graceful small palms, planted in tubs and pots make perfect marginal plantings creating a lush, intimate environment and helping to fill the voids that these areas have by design.

The ceilings of many screened areas sometimes soar to 30 feet or more, and positioning plants can be a problem. I generally use hanging pots of exotics suspended from horizontal wood rails or iron pipes installed across the area like shower curtain rods and fixed to the sides of the enclosure.

Swimming Pool Areas

Swimming pools are a fixture of life in warm climates, and they should be considered part of the landscape plan—with plants and enclosures. Architecturally designed screened housings can enclose a swimming pool, protecting bathers against insects and mitigating the sun's effects. And rarely do you see pools without plantings—a decorative framing that make the area more inviting. Exotic plants in decorative containers makes a sometimes sterile pool area a lush environment. Graceful palms and ferns furnish a tropical ambiance, while plants such as elephant ears or hibiscus in pots and planters offer decorative cooling.

Placing container plants around your pool area requires the same attention to design principles as your general landscape. You want to create a balanced picture without disturbing the exit or entrance to the pool. Do not create a jungle around the pool—consider balance, proportion, and symmetry. In most cases, less is more when embellishing the pool area.

For the most part, plants tolerate some chlorine splash; it is unlikely for foliage to be drenched with pool chemicals, and a few drops of chlorine is harmless to foliage. Avoid plants that drop their leaves: no one wants the continual task of skimming leaves from the water's surface. The same caution holds for fruit trees. Dropping fruit stains the pavement and makes it slippery. Bees and other insects attracted to dropped fruit are a nuisance no bather wants to put up with!

When choosing decorative containers and plants, think big—avoid small potted plants that will be lost against a larger expanse of water. A key design element is to use potted plants as vertical accents to highlight your pool. A palm is an ambitious but dramatic vertical accent. Try the small Arecas, Kentias, and Rhapis for a lush look. Licuala, Livistona and Reinhardtia are other good palms to adorn a pool area. If a palm isn't on your list, many other plants can fill this need. The fine plants listed below will add drama and decoration to your pool area. While some are properly considered water plants (aquatics—see chapter 14), they are all successful container candidates.

POOLSIDE CONTAINER PLANTS

Arrowhead (*Sagittaria*)
Bulrush (*Scirpus*)
Canna (*Canna* x *generalis*)
Cat-tail (*Typha latifolia*)
Egyptian paper plant (*Cyperus papyrus*)
Elephant ears (*Colocasia esculenta*)

This lanai/swimming pool has both outside greenery and, to frame the pool, various plants in clay pots: a Beacarnea, Strelitzia, and a large Cycad. (*J. Kramer*)

A covered lanai, or Florida room, that is accented with a Victorian stand filled with Cattleya orchids and two large *Strelitzia nicolai* in decorative containers. (*J. Kramer*)

A charming poolside scene with Philodendrons; an outside hedge provides privacy. (*J. Kramer*)

Poolside Container Plants *cont.*

Gunnera chilensis
Hibiscus (*Hibiscus*)
Horsetail (*Equisetum hyemale*)
Sweet flag (*Acorus calamus*)
Water canna (*Thalia*)

Ferns

No plants give as much a sense of coolness and softness as ferns. Their graceful beauty makes for a perfect poolside accent, and they are especially attractive in containers. Here are some favorite ferns to try by poolside.

- **Maidenhair ferns (*Adiantum*)**. Fronds are 24 inches long. Somewhat dense, but has a nice, graceful growth habit.
- **Bird's nest fern (*Asplenium nidus*)**. A bright green fern with large fronds. It has a fountainlike growth habit. Very handsome in a decorative pot.
- **Hammock fern (*Blechnum occidentale*).** Dwarfish fern with leathery leaves.
- **Holly fern (*Cyrtomium falcatum*).** Handsome dark green foliage and a graceful manner of growth.
- **Squirrel foot fern (*Davallia bullata*).** Creeping rhizomes clothed with reddish brown scales that resemble a squirrel's paw. Triangular foliage, leathery and bright green.

Six

Container Exotics

Choosing the right container for your exotic pot plants can be bewildering. What looks good in one pot will not necessarily be handsome in another. Airy, dainty plants need bright, bold pots to show them off; coarser, bushy plants need simple containers. A container garden should be pleasing in all aspects—plants, pots, setting—and containers become part of the furnishings. The right plant in the right pot placed in the proper area gives the unity and balance needed to make an outdoor living area beautiful.

Pots and Tubs

The standard unglazed clay pot was one of the earliest planting containers to be used, and it is still very popular. Clay pots come in sizes from 3 to 24 inches in diameter. Because air passes through it, the soil in a clay pot dries out quickly, so there is little chance of overwatering plants. Clay pots are usually inexpensive and durable, and the natural color harmonizes well with indoor and outdoor settings. There are many variations to the standard slope-sided, rimmed pot. Soak new unglazed pots overnight before they are used, or they will absorb water needed by the plants from the soil.

Clay pots run the gamut of origin, quality, and design. The Italian pot modifies the standard border to a tight-lipped detail; it is simple and good-looking. Some have round edges, others are beveled or rimless in sizes from 12 to 24 inches. Venetian pots are barrel-shaped with a concentric band design pressed into the sides. Somewhat formal in appearance, they come in 8- to 12-inch sizes.

Spanish pots are graceful and always charming, with outward sloping sides and flared lips in 8- to 12-inch sizes. They have heavier walls than conventional pots and make good general containers for many plants.

Bulb pans or seed bowls are generally less than half as high as they are wide, like deep saucers with drain holes. They are available in 6- to 12-inch sizes. The azalea pot or fern pot is a squat clay container 6 to 14 inches in diameter. It is three quarters as high as it is wide and is in better proportion to most plants than conventional pots. Three-legged pots are bowl- or kettle-shaped. By raising the plant slightly these containers show smaller plants to good advantage. They range in size from 8 to 20 inches.

The cylindrical terra-cotta pot is a recent innovation, a handsome departure from the traditional tapered design. At present, it is available in three sizes with a maximum 14-inch diameter. Terra-cotta strawberry jars and novelty pots are good when used as accents.

Although unglazed pots are most popular, colorful glazed containers and glass jars are attractive, too. Most lack drainage holes, so water plants in them moderately, because it is difficult to know when the bottom of the pot is filled with waterlogged soil that could kill plants. A 2-inch layer of small stones in the bottom will aid drainage and prevent the soil from souring. If you use glazed pots, take them to a glass store and have drain holes drilled or use them as a decorative outer pot for an unglazed, drained pot.

Plastic pots are lightweight, inexpensive, and come in many colors in round or square shapes. They are easy to clean and hold water longer than clay pots, so plants in them require less frequent watering—an advantage to some gardeners. They are not suitable for large plants because they have a tendency to tip over.

Tubs may be round, square, or hexagonal. If wooden, be sure they are made of durable wood such as redwood or cypress to resist decay. Stone or concrete tubs are ornamental, add dimension to a patio, and are perfect for exotics. Smaller tubs or tapered bowls are especially pretty filled with ginger or crotons.

Japanese soy tubs are inexpensive, handsome, and plants look good in them. They are found at nurseries and basket shops. Wood and bamboo tubs with large foliage plants are also effective outdoors, as are galvanized washtubs painted dark colors.

Sawed-off wine casks banded with galvanized iron are unique and make fine containers. They are commonly made in two sizes, 20- and 26-inch. Barrels and kegs come in many sizes, from small 12-inch diameters to large 24-inch; most are very decorative and plants grow well in them.

Urns and Jardinieres

Concrete, stone, or Fiberglas urns come in many different sizes. They are at their best when viewed at eye level, as on the end of a terrace wall. Pedestal urns are handsome, too, but usually large. They can overpower a small area, so use them with discretion. Stone baskets are happy choices for trailing plants like tropical vines.

Glazed ceramic or porcelain Japanese pots are stunning. If you have a special place that needs rich color, try a blue-glazed porcelain pot with an irregular-shaped shrub such as an Indian hawthorn. Blue- or green-glazed Chinese ceramic pots in various round shapes are handsome and ideal for foliage plants like caladiums and dieffenbachias, and brass pots and gold-leaf tubs make any common plant appear special.

Boxes

Trees and shrubs demand wooden boxes. The largest tub simply does not hold enough soil or carry enough visual weight to balance a tree. Some boxes are simply a perfect cube or a low cube, but once you have the basic container there is much that can be done to vary its design. There are a number of variations for boxes; let your imagination ramble. Dress them up.

A large box with a potted tree weighs several hundred pounds. If you are not in a year-round temperate climate you will have to store the tree in winter. Buy commercial dollies put them under the boxes so they can be moved about easily. Or make your own moving devices with 2-inch casters and boards.

One simple decorative container can create a handsome sight at a doorway; this holds a *Euphorbia millii* (crown of thorns). (A. R. *Addkison*)

A pair of Spanish clay pots bring color to this swimming pool. Three small palms decorate the area. (A. R. *Addkison*)

Echinocactus grusoni, the golden barrel, makes a statement in this decorative container on a lanai. A Chinese Foo dog guards the scene. (A. R. *Addkison*)

This assortment of decorative containers at a nursery is bewildering. In the background is statuary for garden decoration. (A. R. *Addkison*)

Planting Large Containers

Large plants require large containers, and potting and moving big plants can be difficult unless you follow some common-sense guidelines. First, and most simply, decide where to place large containers in your landscape plan or in your pool area. Sketch your site with circular or shaded areas where you want plants. Try different arrangements before deciding where you want your plant. Then, using a hand truck, position the *empty* container where it is to be planted. Do the potting at the desired area rather than potting the plant first and moving it to the site.

Once the containers are in place stand back to observe the total picture: is it symmetrical? attractive? If not, reposition with the hand truck. Or you might want to put plant dollies under the potted containers; then it is easy to move potted plants about if you desire.

Repotting Plants

Whatever growing medium you use, it will have to be replenished in time. Usually plants in containers over 16 inches in diameter can go two or three years without repotting. Smaller containers need fresh soil more often. Repot a plant when you see roots growing out of drainage holes, when the rootball has been compacted, due to lack of water, or you see roots growing through the surface of the soil. Spring and fall are the best times to repot plants. The spring warm weather stimulates plant growth, and in the fall there plants have time to adjust to the shock of repotting before cold weather starts.

Plants that like to be rootbound, such as Clivias and some succulents, only need top dressing. Remove the top 2 to 4 inches of soil and replace it with new soil.

For most other plants, however, transplanting them to new, larger pots is necessary. To remove a plant from a small container, you must loosen the rootball so it will slide out of the container. With a newly purchased plant from a nursery (usually sold in metal or plastic containers), slit the sides of the can with a can cutter (most nurseries will do this for you) before trying to remove the plant. Gently tap the sides of the pot with a hammer. Place the plant on its side, grasp the crown of the plant, and try to ease the plant from the pot.

Wash the new container, cover the bottom with shards of pottery or small pebbles, and insert a 3- to 4-inch bed of soil. (Don't use a pot that isn't large enough to hold enough soil.) Center the plant on the soil

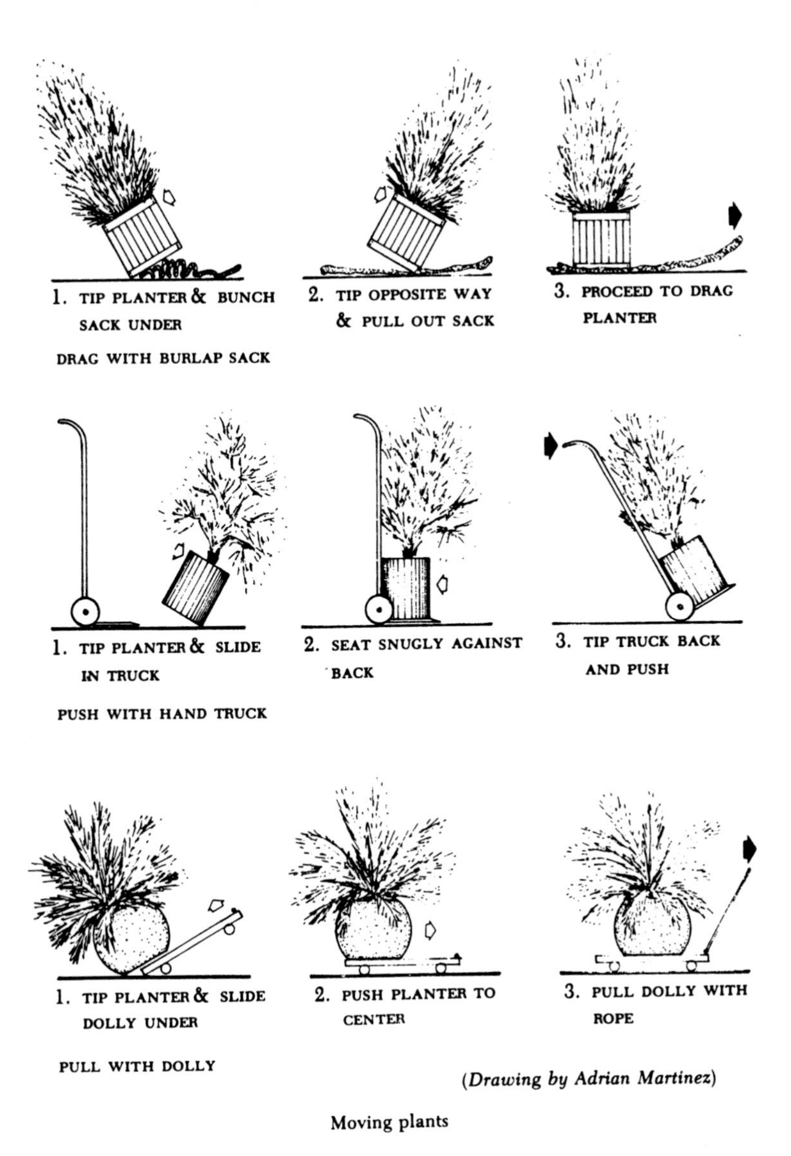

Moving plants

bed and fill in and around the roots with fresh soil. Press down gently to eliminate air gaps. Fill the pot up to 1 to 2 inches below the rim. Soak the soil generously; allow the water to drain, and soak again.

For balled-and-burlapped shrubs and trees, cradle the rootball in your hands and set it in place in the center of the container. Fill in and around the soil up to 1 to 3 inches below the top of the container. Soak thoroughly, allow to drain, then soak again.

Care of Container Plants

Plants in containers constantly take nutrients from the soil and frequent watering leaches out food faster than we realize. Proper feeding keeps plants growing rather than merely living. But often people are anxious to see rapid growth and feed plants too much. Fertilizing plants is beneficial, but too much can harm a plant.

Use a standard 20-20-10 plant food on a regular schedule. For larger pots, feed twice a month during growing times. Plants in smaller containers require feeding more often, perhaps every ten days or so. Feed less in winter and dormant times.

There is no need to feed newly potted plants for several weeks, since there are adequate nutrients in the fresh soil. Ailing and resting plants should not be fed; their roots cannot absorb nutrients and will burn.

500 Beautiful Exotic Plants

Seven

Aroids

At one time or another we all have probably grown some aroids in our gardens or as houseplants. Philodendrons are a prime example, as are spathiphyllums or Chinese evergreens. There are dozens of other plants in the Aroid or Arum family (technically called Araceae) that are desirable for their foliage: some 2500 species are distributed throughout tropical Asia and the Americas. Popular members include: agloanema, pothos, syngonium, homalomena, colocasia, and anthuriums—with more than 700 species.

Aroids are prime candidates for southern gardens because plants grow rapidly and easily and provide a lush green background. Because it is a large family with so many plants suitable for our climate it is difficult to single out every species that could be grown; these are some of my favorites.

Alocasias

Alocasias are exotic plants from Borneo, Australia, and southeastern Asia. A well-grown plant with its heart-shaped leaves and prominent veins is a dramatic sight. These plants are especially good en masse as background plants where you need a real show stopper.

Although most alocasias grow to 30 inches, smaller dwarf types have been introduced recently. These fine plants are best grown in peat moss, perlite and some humus. Good drainage is vital. They require constant moisture at the roots at all times and revel in warmth, minimum 62 degrees F at night. Keep them out of the sun but give them some bright light. Provide 30 to 50 percent humidity and wash leaves occasionally with a damp cloth. At times dormancy does occur, but plants recover.

Still somewhat expensive, alocasias are now available at florists' shops. Here are some I have grown successfully indoors.

- A. *Amazonica* — Handsome hybrid, dark green foliage with white veins.
- A. *cuprea* — Waxy maroon-purple leaves with prominent veins, compact growth.
- A. *Johnstonii* — Foliage marked with red veins.
- A. *lowii grandis* — Deep brownish-green leaves.
- A. *lowii veitchii* — Arrow-shaped leaves mottled.
- A. *Sedenii* — Hybrid with olive green and silver foliage.
- A. *waveriana* — Dark purple lance leaves, purple edges.
- A. *zebrina* — Rare; mottled stems.

Anthuriums

Anthuriums produce red spathlike bracts, waxy and beautiful. Originally red was the only color but through hybridization pink, white, and even some green tints are available. *Anthurium scherzianum* is the most popular species with many new varieties. A. *andreanum* is also seen frequently as a cut flower.

Although most anthuriums grow erect, there are pendent types such as A. *warocqueanum*, and birds' nest anthuriums such as A. *salvinae* are occasionally seen. Some anthuriums can be confused with the larger self-heading philodendrons, but in either case they are lush green accents certainly worth space in the tropical garden.

Caladiums

Caladiums are spectacular in full summer leaf. Leaf color ranges through shades of pink, red, white, or green marked with deeply etched veins of contrasting color. The foliage has a papery texture, and the leaves are

Alocasia cuprea, with its typical arrow-shaped leaves, grows in clumps and adds an exotic touch to gardens. (*John Kilmer*)

The exotic and beautiful *Alocasia amazonica* is a special accent plant for a shady place in the garden. (*John Kilmer*)

Marantas now are available in many varieties; all provide dramatic color and are excellent foliage plants for a shady area. (*John Kilmer*)

Anthuriums are associated with Hawaii but are not indigenous to those islands. They are tropical plants that like warmth and dappled sun; they thrive in the Florida environment, where they bear the popular red spathe (flower). (*J. Kramer*)

Caladiums have been popular garden subjects for years and have a tropical ambiance suitable for southern climes. There are hundreds of varieties. (*John Kilmer*)

Spathiphyllum clevelandii has been a favorite indoor plant for decades, but it can also be used outdoors in Florida gardens. This variety is 'Mauna Lau'. (*J. Kramer*)

pleasingly arrow-shaped. In their native land, the jungles of Columbia, caladiums evolved to distinct wet and dry seasons.

You can start tubers in spring for summer show or plant them in early fall for winter decoration. Start plants in equal parts of sandy loam and peat moss. Keep them warm (minimum 70 degrees F) and place them in bright light. Water them scantily for the first few weeks. Once growth starts, increase moisture and feed them moderately every two weeks. When caladiums start their dormancy, decrease moisture and store the tubers in a warm dry place until the following spring. This is just a sampling of some popular caladium varieties.

- 'Ace of Spades' Lance leaf, red veins over rose and white marbling.
- 'Appleblossom' Transparent apple-blossom pink foliage, green border.
- 'Bleeding Heart' Heart-shaped, pointed leaves marbled white, green, and red.
- 'Caro Nome' Ivory leaves with red veins and overlay of moss green.
- 'Debutante' White and green leaves with rosy central area.
- 'Edith Mead' Small dark green and white leaves, red veins.
- 'Fire Chief' Large pink leaves mottled with dark green.
- 'Gail Dee' Pink, mottled light and dark green.
- 'Jacqueline Gireaud' Yellow-green leaves splashed with crimson.
- 'Kathleen' Pointed rose-colored leaves, dark green edges.
- 'Lord Derby' Quilted rose-pink leaves with green veins.
- 'Maid of Orleans' Deep green, marbled pink.
- 'Miss Chicago' Rose and white leaves with dark green margin.
- 'Pink Blush' Crinkled carmine leaves, thin dark green border.
- 'Pink Radiance' Curled rose-pink leaves with red veins, green border.
- 'Queens Delight' Creamy white and pink leaves.
- 'Red Chief' Lance leaf, red with green border.
- 'Rising Sun' Brilliant crimson, blotched with yellow.
- 'Southern Belle' Rose pink leaves, green edges.
- 'Twinkles' Bronzy red and green.
- 'White Christmas' Pure white with deep green veins.

Dieffenbachia

Dieffenbachias first appeared as houseplants in 1830 in Germany. Yet it is only recently with jet transportation and new growing techniques that we have had a selection of them. Their large heart-shaped leaves resemble caladiums but are not as colorful. Generally, the plants are bushy, lush, and when used indoors are perhaps best in front of windows on the floor rather than on sills. They grow from a central core much like a palm, and although they can become rampant with numerous stalks, pruning makes them lovely.

Coming from the jungles of Central and South America, dieffenbachias are a successful exotic in Florida landscapes. Indoors, they need constant warmth but grow equally well whether in shade or bright light. Pot them in a peatmoss, loam, and sand mixture and add some rotted cow manure. Keep the plants well watered in summer, but carry them somewhat dry (the soil just barely moist) in winter. They are not fussy about humidity and can be used outdoors or indoors.

Keep your pets and children away from dieffenbachia; it is poisonous and its common name, dumb cane, is derived from the fact that chewing leaf or stem brings paralyzing results to mouth and throat.

- *D. arvida* — White leaf pattern.
- *D. amoena* — Deep green leaves blotched white.
- *D. exotica* — Mottled green and white foliage, compact growth.
- *D. picta* 'Rudolph Roehrs' — Arching leaves in shades of green.
- *D. Bowmannii* — Chartreuse foliage.
- *D. Hoffmanni* — Showy white leaf pattern.
- *D. oerstedii* 'Variegata' — Dark green leaf with white midrib.
- *D. splendens* — Velvety green foliage with white dots.

Marantas and Calatheas

The marantas are decorative plants from tropical parts of the world and, although the family is not large (only about 200 species), it is confusing. The group also includes calatheas and ctenanthes. No matter how they are listed in the catalogues, and it will vary, these are superlative foliage plants. The leaves have a crinkly texture and are beautifully marked in olive green or brown with undersides of magenta purple. From the jungles of South America, these fine plants thrive in a shady place; however, do not put them in total darkness and expect them to live. Keep the soil a bit on the wet side during the summer, evenly moist the rest of the year. The plants are at home in warmth, never less than 58 degrees F at night and require 30 to 50 percent humidity. Give them a porous soil of two parts peatmoss, one part sand, and one part loam. A few species go dormant in winter and then require a moist soil.

You may have to search for a few of the species described here. Only popular ones are being offered by nurseries, but I'm sure we will see more of them in the near future.

• C. *argyrala*	Silver gray and green.
• C. *concinna*	Dark green leaves with feather design.
• C. *insignis*	Light green with olive green markings.
• C. *lietzei*	Light green feather design, purple underneath.
• C. *makoyana*	Outstanding olive green, pink, silver, and green foliage.
• C. *ornata*	Also called *Rosea lineata*. Pink on white stripes.
• C. *ornata* var. *sanderiana*	Dark waxy foliage marked with shades of green.
• C. *Vietchiana*	Iridescent leaves.
• Maranta bicolor	Flatter design; dark green, grayish and green.
• M. *Erythreneura*	Bright red veins on pearl gray background.
• M. *Leuconeura massangeana*	Pearl-gray-green foliage.
• *Ctenanthe* 'Burle Marx'	Gray green, dark green leaves.
• *Ctenanthe* 'Oppenheimiana'	Silver and green.

Agloanema

Tough to beat, these plants from tropical forests will thrive under conditions that would kill most houseplants. Outdoors, they do best in containers. Most species have dark green leaves; some have leaves marked with silver or white. Flowers are white. Give them bright light or semi-shade and constant moisture at the roots, or grow them in a vase of water. Agloanema are easy to propagate: Cut stems into 3-inch pieces, place them in moist sand, and barely cover. Only potbound plants will bloom, but bloom they do—luxuriantly—in late summer and early fall.

A. *commutatum*, growing to 24 inches, has silver-marked, dark green leaves. A. *commutatum* 'Pseudo-bracteatum', generally classed as a hybrid, grows to 24 inches; it has green leaves splashed with yellow. A. *modestum*, which grows to 24 inches, has waxy dark green leaves. A. *pictum*, at only 12 inches, has dark green, velvety leaves spotted with silver. A. *roebelinii*, which may reach 36 inches, has blue-green leaves. It is very robust. A. *simplex*, another tall variety growing to 36 inches, has dark green leaves. A. *treubii*, growing to 24 inches, has lance-shaped, bluegreen leaves marked with silver.

Colocasia

Colocasia have always been among my favorite foliage plants, with their elegant, large, velvety green leaves on tall stems. Keep these plants in a bright place and water them heavily during active growth. In winter, when plants are dormant, keep them almost dry in their pots at 60 degrees F. Grow new plants by division of tubers.

C. *antiquorum illustrus* grows to 48 inches and has green leaves with purple spots. C. *esculenta* reaches 40 inches, with quilted satiny green leaves.

Monstera

Monsteras are huge climbing foliage plants, often reaching 72 inches or more, with 24-inch perforated green leaves, making them excellent for planters in public rooms or outdoors in the garden. Grow them in bright light and keep soil consistently moist. They like 50 percent humidity. They grow in water almost better than in soil. Plants will adjust to warmth or coolness, but keep them protected from frost. Older plants bear unusual boat-shaped white flowers. Aerial roots grow from stem nodes; these can be cut off if you wish, or used to propagate monsteras from stem cuttings or by air layering.

M. *acuminata* is a smaller leaved variety with 14-inch leaves. M. *deliciosa* produces 36-inch leaves with pronounced slits along the edges.

Spathiphyllum

Twenty-inch aroids from South America, spathiphyllums have shiny leaves and white flowers carried in bracts called spathes, resembling anthuriums. Bloom usually appear in winter, but sometimes in summer or fall. Give plants a bright location and keep soil consistently wet in winter; then grow them on the dry side. Propagate them from seed or by root division.

S. *cannaefolium*, which grows to 24 inches, has leathery black green leaves; S. *clevelandii*, which grows to 20 inches, has long leaves; and S. *commutatum*, reaching 30 inches, has broad, elliptical green leaves. S. *floribundum*, a dwarf that grows only 14 inches tall, has green leaves and white flowers. S. *floribundum* 'Marion Wagner', which reaches 30 inches, has quilted, rich green leaves, and S. floribundum 'Mauna Loa', which also grows to 30 inches, has satiny dark green leaves.

Philodendron

In the large family of philodendrons (there are hundreds), some species are better than others for decorative accent. Some are straggly plants that add little to the garden, others are stunningly graceful.

Suggesting the best philodendrons is a hazardous task, but through the years I have grown so many I feel I can recommend certain plants that have more outstanding characteristics—leaf shape, easy growth, color—than others.

The philodendrons are climbing shrubs forming groups of 1,220 species native to tropical America. (The word philodendron derives from *philo*, "to love," and *dendron*, "tree.") In their native habitats, many philodendrons clothe trees and spiral to great heights to display their lush foliage. Other species are self-heading—they grow in a rosette fashion and do not climb. In nature philodendrons prefer a slightly moist atmosphere and warmth. The plants bear leathery, oblong, heart-shaped, or lobed leaves and generally are fast growers, especially the vining types. There are many philodendrons, but until recently only a few species were available commercially. Now, many more species are being introduced, and the selection is good.

How do you find the best philodendrons for your garden? Visit local sources and see what they have to offer; then write for mail-order supplier's catalogs (listed in the back of this book). You want a plant that is pleasing to look at and that will perform satisfactorily in the environment you can give it.

Vining. The viners are, by nature, epiphytes—their roots and leaves derive moisture from the air. Vine-type philodendrons need a support. In the vine or climbing group there are cut-leaf species, such as *P. radiatum* and *P. panduriforme*, or solid-leaf types, such as the popular *P. hastatum*, *P. cordatum* (oxycardium), and *P. andreanum*. Leaf color varies in both groups: Some are light green, some are dark green, and others, such as *P. andreanum*, are almost blackish green.

Vining Philodendrons

- *P. andreanum* — Iridescent, velvety, oblong leaves are dark olive green with a copper tinge and ivory-white veining. This handsome plant needs high heat and humidity to thrive.
- *P. asperatum* — With thick stems, and dull green, broadly cordated leaves, this plant grows to 18 inches and is deeply corrugated by numerous sunken veins.
- *P.* 'Burgundy' — A robust wine-red hybrid has lance-shaped leaves.
- *P. cordatum oxycardium* — This popular heart-shaped plant with lustrous green leaves is grown easily with good moisture.
- *P.* 'Emerald Queen' — This popular and lovely hybrid with deep green, medium-sized shiny leaves is a compact grower.
- *P. erubescens* — A climber with waxy, bronzy green foliage that is easy to grow.
- *P. glaucophyllum* — This deeply ribbed plant with striking heart-shaped, greenish-black leaves can become spindly, but is easy to grow.
- *P. hastatum* — This plant with closely set, shiny, arrow-shaped leaves needs good moisture and humidity.
- *P. hastifolium* — With deep green, durable, broad-cordate leaves, it grows to 18 inches when mature.
- *P. imbe* — These long, arrow-shaped leaves are somewhat red and have handsome veining. This fast grower is amenable to most conditions and can grow quite tall.

This is a typical Philodendron. The species is *P. oxycardum*, but there are many varieties, each slightly different in leaf shape. Most Philodendrons make fine plants for somewhat shady places. (*J. Kramer*)

Philodendron selloum (*Barbara J. Coxe*)

Philodendron saggittatum is a popular Philodendron, a climbing plant that affords lush leaves and fast growth in the tropical environment. (*John Kilmer*)

Pothos are known by a variety of names but basically resemble variegated, small-leaved Philodendrons. They do well in southern climates and grow rapidly in the landscape. (*J. Kramer*)

Calathea regalis is a colorful Aroid that can be used as an exotic in many landscapes; it can grow to 10 feet. (*John Kilmer*)

• *P. mamei*	This moisture loving plant's large, grassy green, arrow-shaped, waxy leaves are marbled with silver and create a nice canopy effect.
• *P. mandaianum*	The first philodendron hybrid in the United States has arrow-shaped, dark green leaves, with reddish to wine highlights. Although spindly and not the best, it is easy to grow.
• *P. ornatum*	Its fresh green, leathery leaves have depressed veins.
• *P. panduriforme*	This strong, easy grower has fiddle-shaped, leathery textured, olive-green leaves.
• *P. pertusum*	This form of *Monstera deliciosa*, has small, somewhat round, dark green leaves that are deeply scalloped, and grows quickly. The variegated form is outstanding, but difficult to find.
• *P. radiatum*	This plant with broad, rich green, deeply-lobed leaves can grow big and lush with plenty of moisture.
• *P. rubrum*	These large, lance-shaped, lush green leaves can become spindly, but is easy to grow.
• *P. sagittatum*	A plant that will tolerate some neglect, it has large and leathery oblong leaves with sunken veining.
• *P. sanguineum*	Its lance-shaped, dark green leaves have a light midrib.
• *P. sodiroi*	Its bluish-green, glossy leaves are sometimes marbled with silver. When they grow large, these plants are spectacular. There are two types, one with small leaves and one with large.
• *P. speciosum*	This rare climbing, treelike plant has lush green, mammoth, lance-shaped leaves.
• *P. squamiferum*	Large, glossy green leaves are lobed in five sections. This good, slow viner is lush, but is intolerant of drafts.
• *P. verrucosum*	These heart-shaped, dark bronzy green leaves have margins of emerald green, and are sometimes purple underneath. This colorful plant likes warmth and humidity.

Self-heading. If the term "self-heading" confuses you (and it does me), just think of plants that grow in a rosette or cabbage shape. These are the lesser-known species of philodendrons, but to my eye they are as beautiful as the vines. Self-headers are compact, generally large plants from 2 to 4 feet across, and make fine accents for gardens.

Like the viners, some self-heading species have cut leaves, such as *P. selloum* and *P. bipinnatifidum*; others have solid leaves, such as *P. cannifolium* and *P. wendlandii*. Aerial roots are absent in the rosette group, and the plants are slow growers. Like most philodendrons, however, plants grow readily with reasonable care.

Self-heading Philodendrons

- *P. bipinnatifidum* — The skeletal appearance of this plant is dramatic, with its leathery and lobed metallic green leaves and depressed veining. Give it even moisture.
- *P. cannifolium* — With its large-lance-shaped leaves and swollen leaf stalks, this is a nice, erect, low-growing plant. A slow grower, it needs good humidity.
- *P. corcovadensis* — Its tough and leathery arrow-shaped leaves make a nice, compact rosette. Keep it on the dry side.
- *P. eichleri* — These magnificent, huge, metallic green pendant leaves have scalloped edges.
- *P. fragrantissimum* — Large, broad, glossy green leaves are corrugated with many depressed veins. This plant likes even moisture and heat.
- *P. longistilum* — Glossy green, leathery, almost spatula-shaped leaves have wine-red backs and make a colorful accent.
- *P.* 'Lynette' — This plant has a bird's nest shape with leathery apple-green leaves and ribbed depressions. It makes a fine, compact rosette and likes moisture.
- *P. poeppigii* — These cupped, heart-shaped, glossy green leaves with slightly curled edges are stellar. Keep this plant evenly moist.
- *P. selloum* — This plant of very large, dark green, pendant-shaped leaves with a short lobe at the tip is fast-growing, lush, and massive. Likes moist conditions.

- *P. undulatum* — The somewhat small, waxy, lobed leaves, are generally cupped and carried on erect stalks. Seek mature plants that are lush and handsome. Provide good moisture and humidity.
- *P. wendlandii* — This large plant, with broad, waxy green leaves that have thick midribs, makes a stunning compact rosette. Variegated form is outstanding but hard to find. Allow it to dry between waterings.
- *P. williamsii* — These shiny, dark green, very large, arrow-shaped leaves form a good-looking, yet somewhat tall, rosette.

Eight

TREES

When one thinks of trees in Florida, palms immediately come to mind—but think again. Palms do grow luxuriantly in Florida, but so do many other trees of great beauty for gardens. Trees are the skeleton of a garden, and choosing correctly can be difficult. A good tree is one that tolerates Florida's climatic conditions without much care, provides a vertical accent to balance other plants in the garden, and is strong enough to withstand tropical winds and, heaven forbid, hurricanes.

Once these basic criteria are met, select trees that, when mature, will have a pleasing shape and stay in proportion to the landscape. Avoid trees that shed lots of leaves or fruit, because clean-up will be continuous. Because we generally buy trees in 5- or 10-gallon cans or balled-and-burlapped, it is difficult to visualize the same tree in five or ten years.

A landscape contractor or architect can help you choose and place trees wisely, but be sure to select with care. Many landscape maintenance people promote themselves as landscape experts but may not be qualified. Always ask to see credentials and ask for references. You may choose to hire professional services by the hour, hire a professional to supply a sketch or plan, or hire a company to do the complete job from planning to planting.

Planting a Tree

Digging holes for trees is tough work, so, if possible, hire it out. Holes for trees must be deep and large because the roots will spread out. Dig holes at least 30 to 40 inches deep and 5 feet wide. Break up existing soil and add some humus or topsoil. Untie burlapped trees, but leave the burlap in the ground to rot. Canned plants must be removed with a can cutter (most nurseries will do this). People often make the mistake of burying too much of the trunk below the ground. Set the tree so that the soil level is almost the same as it was in the can or burlap.

It is not necessary to prune trees at planting time. Some people do trim bare root trees claiming that trimming makes them grow more vigorously, but I have never noticed a difference in the rate of growth between a pruned tree and an unpruned one.

Canned trees may be planted at almost any time but bare-root or balled trees do best when planted in summer. When any young tree is planted it should have suitable support to withstand wind during its young growth. Guy wires are placed at even distances around the tree; bend and loop each wire and thread it through a rubber hose that will go around the trunk at the point of attachment. Fasten the guy wires in place and attach them to 24 inch stakes set firmly in the ground and placed at equal distances around the tree. Pull the guy wires taut to hold the tree in place.

After planting, water the tree thoroughly; allow water to seep into the soil and then water again. With all plants, create a deep saucerlike depression around the base of the tree so that ample water can get to the roots and not spill off.

Table 8-1 lists some fine trees for the Florida landscape.

TABLE 8-1

Botanical & Common Name	Mature Height	Min. Night Temperature, °F	Remarks
Acacia longifolia	30'	30 to 40	Evergreen tree; likes dry soil (many varieties)
Acer rubrum (red maple)	50–75'	20 to 30	Popular tree
Albizzia julibrissin (silk tree)	20'	5 to 10	Delicate foliage

Botanical & Common Name	Mature Height	Min. Night Temperature, °F	Remarks
Araucana araucana (monkey puzzle tree)	90'	5 to 10	Beautiful open form
Bauhinia blakeana (orchid tree)	30'	20 to 30	Showy flowers
Brachychiton bubuineus (flame bottle tree)	20'	30 to 40	Showy flowers
Brousonetia papyrifera (paper mulberry)	30'	5 to 10	Dense round-headed tree
Camellia japonica	45'	5 to 10	Many varieties
C. *sasanqua*	20'	5 to 10	Many varieties
Cassia corymbosa (flowering senna)	10'	20 to 30	Free-flowering
Cercis canadensis (Eastern redbud)	40'	-10 to -5	Lovely dark pink flowers
Cinnamonum camphora (camphor tree)	40'	20 to 30	Very dense evergreen tree
Citrus sp.	20–30'	30 to 40	Many varieties, all fragrant bloomers
Clusia rosea	30'	30 to 40	Shade and flowering tree
Cordia sebestena (geiger tree)	25'	30 to 40	High drought tolerance
Couroupita guaianensis	25'	30 to 40	Exquisite orange flowers
Crataegus (hawthorn)	20–40'	20 to 10	Excellent color
Delonix regia (poinciana)	30'	30 to 40	Outstanding flowers
Diospyros ebenasta (persimmon)	45'	10 to 20	Resistant to pests
Eriobotrya japonica (Japanese loquat)	15'	5 to 10	Easy to grow
Eugenia uniflora (Surinam cherry)	25'	30 to 40	Glossy-leaved evergreen
Feijoa sellowiana (pineapple guava)	20'	10 to 20	Edible fruits
Ficus benghalensis	85'	20 to 30	Sandy soil

Botanical & Common Name	Mature Height	Min. Night Temperature, °F	Remarks
Grevillea robusta (silk oak grevillea)	50'	30 to 40	Sandy soil
Hymenosporum flavum (sweet shade tree)	50'	30 to 40	Rapidly growing evergreen tree
Jacaranda acutifolia	30'	20 to 30	Large blue-violet flowers
Koelreuteria formosana	30'	20 to 30	Flat-topped tree
Lagerstroemia indica (crepe myrtle)	15'	30 to 40	Luxuriant flower plumes
Leptospermum aevigatum (Australian tea tree)	25'	20 to 30	Evergreen tree for poor soil
Liquidambar formosana (sweet gum)	100'	5 to 10	Good autumn color
Lirodendron tulipifera (tulip tree)	75'	20 to 10	Pyramidal, massive tree
Macadamia ternifolia (Queensland nut)	35'	30 to 40	Slow-growing evergreen
Magnolia grandiflora (Southern magnolia)	60'	5 to 10	Pyramidal tree
M. *virginiana* (sweetbay magnolia)	60'	5 to 10	Big, waxy white flowers
Mangifera indica (mango)	90'	20 to 30	Sweet fruit
Melia azedarach (chinaberry tree)	45'	5 to 10	Round-headed, branching tree
Myrica cerifera (bayberry)	30'	5 to 10	Slender, upright shrub or small tree
Myrtus communis (myrtle)	10'	20 to 30	Good shrub or small tree for hot, dry spots
Nandina domestica (heavenly bamboo)	8'	5 to 10	Fresh and green in droughts
Nyssa aquatica (water tupelo)	50'	5 to 10	Outstanding for watery spots
Osmanthus fortunei (osmanthus)	12'	10 to 20	Vigorous; superior

A favorite Florida tree is *Tabebuia caraiba*: it is generally deciduous with butter-yellow flowers that can be seen from miles away. (*A. R. Addkison*)

Tabebuia blossom. (*Barbara J. Coxe*)

(*Barbara J. Coxe*)

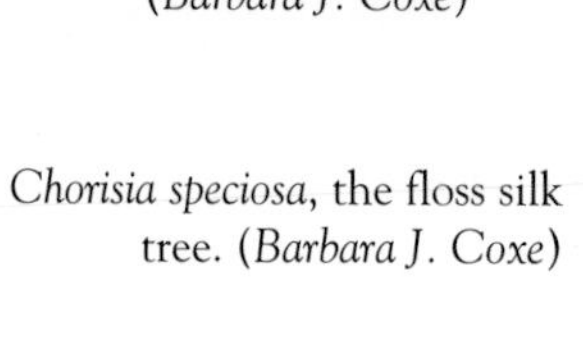

Chorisia speciosa, the floss silk tree. (*Barbara J. Coxe*)

Couroupita guaianensis is a botanical curiosity because it changes its foliage color and form of fruits and flower very unusual. Orange flowers come from branches. (*A. R. Addkison*)

Plumeria alba, the popular frangipani tree, is low-growing and branching. Plumeria makes a fine accent in warm, sunny climates and is highly recommended. (*A. R. Addkison*)

Bauhinia blakeana, the orchid tree, is frequently seen in Florida landscapes. There are many species, but trees tend to be low-growing and messy. (*A. R. Addkison*)

Frangipani. (*Barbara J. Coxe*)

Botanical & Common Name	Mature Height	Min. Night Temperature, °F	Remarks
Pandanus utilis (screw pine)	20'	30 to 40	Easy to grow
Photinia serrulate (Chinese photinia)	30'	5 to 10	Red berries
Pinus palustris (long-leaf pine)	80'	10 to 20	Pretty, round-headed tree; easy to grow
Pistacia chinensis	50'	20 to 30	Good Florida tree
Platanus occidentalis (oriental plane tree)	90'	-5 to 5	Maplelike foliage
Plumeria alba (frangipani)	15'	30 to 40	Fine flowers
Podocarpus macrophylla (yew podocarpus)	60'	5 to 10	Dense evergreen tree
Poncirus trifoliate (hardy orange)	35'	10 to 20	Good hedge
Prunus caroliana (Caroline cherry laurel)	30'	10 to 20	Glossy leaves
Psidium littorale (Guava)	25'	30 to 40	Grown for fruit
Spathodea campanulata (African tulip tree)	25'	30 to 40	Fast grower, showy flowers
Tabebuia caraiba (gold tree)	20'	30 to 40	Fine yellow flowers

Favorite Trees

The many unusual tropical trees that can be grown in Florida must be seen to be believed. Across from my house is a gigantic poinciana whose orange halo literally stops traffic in the summer. And speaking of dramatic colors, no one can resist the stunning purple-violet flowers of the jacarandas. The lipstick tree, the flame tree, the orchid tree—all provide magnificent color and superb beauty.

The acacia, with its canopy of lovely yellow flowers, is fast growing and well suited to landscapes in the central and lower southern parts of Florida. It grows to about 30 feet, tolerates salty soil, and can even survive droughts if necessary. Grow it in zones 9 to 11.

The avocado (*Persea americana*), a favorite tree in Florida, has a strong branching habit and thus is a good shade tree. However, the plant does require excellent drainage to do well. Several varieties are available; buy only those proven in your area. This fine specimen tree does well in zones 9 to 11.

Bauhinia blakeana, or orchid tree, is spectacular in bloom, with masses of orchidlike, deep wine-red flowers that last about four months. The tree has an umbrellalike crown and grows to about 25 feet; it will tolerate salty soil and makes a good specimen or accent plant. Grows in zones 10 and 11.

The bottlebrush trees *Callistemon rigidus* and *C. virginalis* are fast growing and display profuse masses of red flowers in plumelike growth. These trees grow to 25 feet in height and can tolerate a somewhat sandy soil. Its weeping growth makes it desirable in the landscape as both a specimen plant and a good background tree. Grow it in zones 9 to 11.

Chorisia speciosa, the floss silk tree, is not usually seen in Florida, but it is a fine accent plant. It grows to about 35 feet with a nice canopy. It is quite pretty when it blooms, generally in the autumn. It can sometimes go deciduous and usually adapts to conditions in the entire state. Grow it in zones 9 to 11.

The Japanese loquat, *Eriobotyrya japonica*, is the tree that produces apricotlike fruit no one seems to know what to do with. Actually, the heavily seeded fruit is great when cooked with chicken, pork, and other foods. The tree is full and compact and displays white flowers. It grows to only about 15 feet, so it is a border tree. Grow in zones 8 through 11.

Brachychiton popuineus, or flame tree, likes a hot climate. This compact tree grows to about 20 feet and has rich dark green leaves and bright red flowers. Several varieties are available. Grow the flame tree in zones 9 to 11.

A spectacle in bloom, *Jacaranda mimosfolia* and *J. acutifolia* both bear dramatic violet-blue flowers. Generally evergreen, these trees grow to about 30 feet and have a fine spreading habit. The jacarandas take to the Florida heat but are somewhat sensitive to the cold, although I have seen them growing as far north as zone 6.

Lagerstroemia indica, the crepe myrtle, is sometimes classified as a shrub, but in southern Florida it grows into a tree about 15 feet tall. Various varieties are available; the plumes of pink to rose flowers are stunning in bloom. The crepe myrtle takes to warmth. Grow this good accent tree in zones 9 to 11.

The stunning poinciana. (*Barbara J. Coxe*)

Japanese loquat. (*Barbara J. Coxe*)

Weeping bottlebrush. (*Barbara J. Coxe*)

The African tulip tree, *Spathodea campanulata*, grows to about 25 feet. The great color lasts quite a while; one in my neighbor's yard displays color six months of the year. This unique tree does well in southern parts of the state (zone 11) but may grow in zones 9 and 10 as well.

Of all the Florida trees, the poinciana (*Delonix regia*) is the most stunning. Its beauty is so majestic that songs have been written about it. With its ferny foliage and umbrellalike growth, the poinciana can grow to about 40 feet. The tree displays its beautiful color in the summer; it grows best in zones 10 and 11.

The gold tree, Tabebuia, is the one I am most often asked about, and rightly so: With its masses of bright yellow flowers, it is a striking sight, quite dramatic. It generally grows to about 20 feet and is a good accent tree. Grow in zones 10 and 11.

I often wonder why I do not see more *Plumerias* grown in southern Florida because they do put on an impressive show with their white or pink flowers. More commonly called *frangipani*, the tree usually is tall, with forking branches. Unfortunately, frangipanis are sensitive to cold and so must be protected when temperatures reach 45 degrees F or below. Grow in zone 11.

Some trees, such as magnolias, redbuds, and dogwoods are not considered exotic but offer great color and beauty to our Florida landscapes and are used frequently in garden designs, especially in the northern tier of Florida near Sarasota and Gainesville.

Nine

Palms and Cycads

Palms are Florida's trademark, and rightly so, for they are graceful, tropical, and majestic. The same varieties—Areca, Arecastrum, and a few others—are used constantly, yet there are many other palms equally as useful and attractive.

Most Florida nurseries, garden shops, and even large supermarkets sell palms, but wherever you buy the plants, look for healthy, vigorous specimens with full green crowns. Most importantly, get the proper botanical name; too often places such as supermarkets call everything "areca palm" or just "palm." Knowing the correct name of your palm means that you can give it the proper care.

(If you want an instant landscape, buy full-grown palms from a professional grower or landscape firm. Note, however, that mature palms are very top heavy and hard to maneuver; they must be planted by professionals.)

Only a few palms are cold hardy, a fact you must consider if you live in northern Florida, which sometimes gets hit with freezing weather. Palms have a shallow root system, so they need bracing or guying to encourage proper growth. Guy wires are available at nurseries; attach them to stakes placed in the ground in a triangular pattern. Leave the bracing in place for about a year.

Palms need a careful feeding program and fairly good soil to thrive. In southern Florida, where soil is sandy, add loam and peatmoss. Four times a year, feed plants with a palm fertilizer: Twice in the spring, once in the summer, and once in the autumn.

Palms naturally lose their bottom fronds after a time. Because most trees are so tall, have professionals do the necessary trimming and pruning. Trim old fronds while they are yellow rather than waiting until they turn brown; they are difficult to remove when they are brown.

Palms can be transplanted at almost any time of the year, but I prefer to move them just before the rainy season arrives. This way the plants have a few weeks to become acclimated to their new location. Thereafter, rain will supply the water the plants need for the development of their roots. Palms usually respond quite well to being moved and take root with no problems.

Plant palms deeply; shallow planting can injure them. Once the palm is in its deep planting hole, fill in with fertile loam, and let water flow in to remove air pockets. Tie the leaf stems up around the center of the palm, to protect the leaf bud from injury. Palms bear only one growing point; if the bud is bruised or broken, the palm will not survive.

Growing Palms

Florida is called the Sunshine State, but it is really the palm state because so many dot the landscape as you drive south. For the past decade I have been working with palms—planting them, supervising the cutting of plants, moving the plants, and caring for them—and I have discovered several easy maintenance tips not covered in any gardening books about Florida.

Use palms suitable for your zone. Some palms may succeed out of their natural zone, but there is no guarantees. Along the coast from Jupiter to Bradenton you can use almost any palm, but as you head north and inland, the selection is not as great. Here are the palms I recommend you try, along with tips for successful growing:

- *Areca catechu*, the betel nut palm, is omnipresent in hot climates, where it grows well and quickly. Old leaves droop and need pruning; otherwise, easy to grow in zones 10 and 11. A smaller version of the betel palm, *A. ipot*, grows to 12 feet and is also suitable for zones 10 and 11. *A. triandra* is an attractive palm with a clumping growth habit. It's a favorite with landscapers.
- *Bismarckia nobilis* is a handsome, tall palm with broad leaves. Great beauty for zones 10 and 11.

Pygmy date palm (*Phoenix roebelenii*). (*Barbara J. Coxe*)

The multiple-trunked *Phoenix canariensis* is a favorite slow-growing Florida palm. There are many varieties, some of which grow to 40 feet. (*A. R. Addkison*)

Arenga engleri is a low-growing palm that is perfect for smaller gardens. Also called the sugar palm, it grows quite large to 25 feet. (*A. R. Addkison*)

Tall and stately, the coconut palm, *Cocos nucifera*, reaches 70 feet with medium growth. (*A. R. Addkison*)

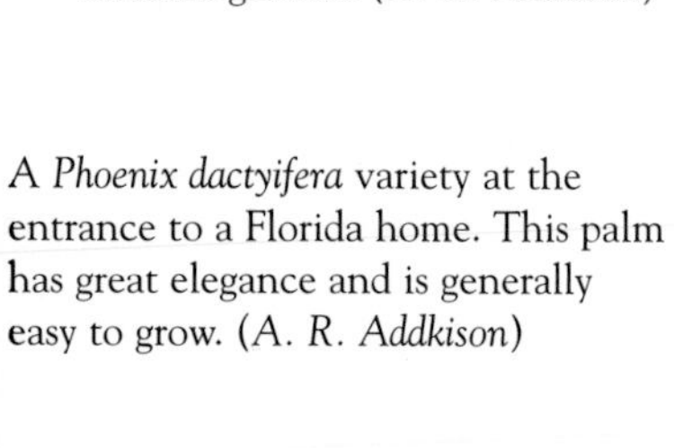

A *Phoenix dactyifera* variety at the entrance to a Florida home. This palm has great elegance and is generally easy to grow. (*A. R. Addkison*)

Bismarckia nobilis has distinctive gray-green fronds of beautiful design; it is an excellent landscape tree. (*A. R. Addkison*)

The fan-shaped palms of *Sabal minor* create a lush effect in this tropical garden. These plants can grow to be quite large. (*A. R. Addkison*)

Cocos palms in the landscape. (*A. R. Addkison*)

Washingtonia robusta is a tall palm that is more suited to street decoration than a private garden. (*A. R. Addkison*)

- The fishtail palm, *Caryota mitis*, is incredibly beautiful when mature, with dark green fronds that look like fishtails. In the northern part of the state the fishtail palm is a houseplant, but farther south it dots the landscape, growing to 15 feet. This palm is very easy to care for; it responds well to summer rains and winter droughts. If trimmed properly, the fishtail palm is a very decorative element against a house wall or in a corner of the garden. Few insects bother this palm. Note that the fishtail palm is sensitive to cold (below 45 degrees F); it does especially well in zones 10 and 11.
- *Chamaedorea elegans*—parlor palm or good luck palm—is attractive but quite sensitive to the cold. It grows about 8 feet tall and makes a nice accent in a small yard. Be sure to protect it from cold. Grow in zones 10 and 11.
- *Chamaedorea erumpens*, the bamboo palm, is not often grown, but it should be because it is so graceful and lush, reaching about 30 feet. It can be grown in most parts of Florida, although it needs some protection in the northern areas. Ideal for zones 9, 10, and 11.
- *Chamerops humilis* (fan palm) tolerates many types of soil and grows well in all parts of Florida. Not overly handsome but nice. Grow in zones 9 through 11.
- Another landscaper's favorite, *Chrysalidocarpus lutescens*, the butterfly or areca palm, grows to about 20 feet—bunching above ground level. A good clumping palm, it is easily grown in full sun or shade. For zones 10 and 11. *C. madagascariensis* is a similar palm, but grows to 30 feet on a solitary trunk. It does, however, require full sun. For zones 10 and 11.
- *Cocos nucifera*, the coconut palm, is probably the most popular palm in the world. It is elegant in stature with a leaning but graceful trunk. The coconut palm is easy to grow and loves humidity. If you're thinking about this palm in your home landscape, take into consideration its tall height—to 80 feet. For zones 9, 10, and 11.
- The sentry palm, *Howea belmoreana*, grows to 25 feet with a slim trunk. It does adapt well to hot climates, but is finicky about conditions and prefers some shade. Still, it has its fans. For zones 9, 10, and 11. I do highly recommend *H. forsteriana*, or the

Kentia palm. This handsome favorite branches to about 50 feet. It likes sun and heat in zones 9 to 11.

- *Jubaea chilensis*, the Chilean wine palm, has a massive trunk and grows to 75 feet. This palm is best used as a street tree. It does moderately well in heat—zones 9 and 10.
- *Livistona chinensis*, the Chinese fan palm, grows to about 20 feet and is ideal for where you want cover rather than height: I use *Livistonas* as hedges. These lovely and decorative plants have graceful fronds and tolerate salt in the soil fairly well. Chinese fan palms will grow in either sun or shade and need only minimal care. Insects and diseases hardly ever bother the Chinese fan palm, which is ideal for zones 10 and 11.
- *Phoenix canariensis* (Canary Island date palm) grows in all parts of Florida and is most desirable. It grows to about 40 feet and has a very nice fountain-shaped growth, making it useful in many situations. On Florida roads the date palm is often used in median strips. Grow in zones 10 and 11. *P. dactyifera*, the date palm, grows to 70 feet, and is generally a street tree.
- I have always been fond of *Phoenix roebelinii*, the date palm, even back when I lived in northern California and grew it as a tub plant on my patio. This palm is not overly handsome, but it grows very easily and reaches about 40 feet when mature. It grows well in zones 9, 10, and 11.
- The lady palm, *Rhapis excelsa*, is tough to beat if you want a low-growing palm that will respond well to the Florida climate. This very attractive clumping plant has fronds of dark green leaves. It rarely grows to more than 12 feet. Though somewhat sensitive to the soil, it can grow in all parts of the state. Fine for zones 10 and 11.
- The stately *Roystonea* species includes many tall-growing trees. They grow well away from Miami to the middle of the state and can reach 80 feet. Because of their size, these giants are best used in large landscapes; they can overpower a small location. *R. regia* is most popular. Grow these in zones 10 and 11.
- *Sabal palmetti* (cabbage palm) is cold hardy and grows easily in all parts of Florida. However, it is not especially handsome; in fact, it is rather ungainly looking. It is useful, however, in areas that receive cold temperatures (zones 9 through 11).

- *Veitchia merrillii*, the Christmas palm, is popular in cultivation. It grows to about 20 feet on a solitary, smooth trunk. It is fast growing and likes water. For zones 9 to 11.
- The Washington palm, *Washingtonia robusta*, has a polelike growth and tolerates salt in the soil fairly well. It is used throughout Florida as a street tree; it looks best when grown in larger landscapes. It grows slowly, and when mature it is a handsome plant. The Washington palm will thrive in all three areas of Florida: upper, middle, and lower (zones 9, 10, and 11).

I would think twice before planting the areca palm (*Chrysalidocarpus lutescens*). Florida abounds with areca palms, and they do tolerate a salty soil, but they are not handsome plants. The regal queen palm *Arescastrum* and the coconut (*Cocos nucifera*) and cabbage palms (*Sabal palmetti*) are better as street and park plants than as landscape subjects in home gardens.

Cycads

Cycads are some of the oldest living plants on Earth and are frequently mistaken for palms. Cycads have a handsome fountainlike growth that emerges from a central core in glossy dark green fronds. They make excellent landscape plants because they are truly tropical in appearance and offer a lush green effect. In general, the cycads require the same type of care as palms. Most species are extremely handsome and can grow into large specimens.

Of the cycad family, I have grown species of *Cycas*, *Dioon*, and *Zamia*. The *Zamia* genus differs from the other two in having larger leaves (fronds). The following are Cycads I recommend for Florida.

Cycas revoluta, known as the Sago palm, is the hardiest of the group and can survive 30 degrees F without harm. *C. circinalis* grows somewhat larger with graceful fronds. *Dioon edule* looks like a squat palm and is very handsome but seldom seen. *Zamia furfuracea*, called the cardboard plant because of its thick leaves, is a spectacular landscape plant often used in commercial plantings. It is my favorite of the group.

Ten

SHRUBS

Shrubs form the background of a good garden and can be used as accents, hedges, or border plantings. Some shrubs are grown primarily for their foliage and others are grown for flowers or even fruit.

Balance and proportion must be maintained when placing shrubs; one or two here and there just won't do, but rather, use shrubs to frame the landscape, perhaps a few on one side of the property and two or three others on the other side. If you are using shrubs as hedges then of course plant them accordingly—in a row to define your property, to afford privacy, or to create a green wall. In this case spacing is important and the general rule is 3 to 5 feet apart depending upon the shrub used. It is important not to plant too many shrubs to avoid an overgrown effect. The plant you buy in a 1- or 5-gallon tub is not the size it will be when it reaches maturity; try to visualize the plant a few years down the road when it is fully grown. Check the plant tag or ask a nursery worker for mature height information.

Be sure to pay attention to leaf texture. A small-leaved shrub next to a large-leaved shrub is not quite attractive to view. Use small-leaved shrubs in one area and large-leaved ones in another. Your garden will have a hodgepodge effect if you try to mix.

It is also important to keep flowering and foliage shrubs in separate areas. If at all possible, group shrubs that need copious watering together, and those that can withstand some drought in another area. The same rules apply to light: sun lovers in one area, shade lovers in another.

Planting Tips

Holes for shrubs should be large and deep because the roots spread out when you put the plants in the ground. For large shrubs (sold balled-and-burlapped, bareroot or in cans) dig holes at least 20 to 30 inches deep. Break up soil in the bottom of the hole and add some humus. Untie burlapped plants but do not disturb the burlap—leave it around the rootball and it will eventually decay. Remove canned plants by using a cutter (or have that done at the nursery). A common mistake is to plant shrubs too deeply; set the plants so that the soil level is almost the same as it was in the can or burlap.

It is not generally necessary to prune shrubs at planting time, but you may want to remove broken or dead twigs and root ends.

TABLE 10-1

Botanical & Common Name	Mature Height	Min. Night Temperature, °F	Remarks
Abelia floribunda	6'	10 to 20	Evergreen shrub
Abutilon hybridum	6'	10 to 20	Evergreen shrub
Acalypha wilkesiana (bronze leaf)	8'	30 to 40	Bronze foliage
Aucuba japonica (Japanese aucuba)	10'	20 to 30	Good leafy shrub, dense
Beleperone guttata (shrimp plant)	8'	30 to 40	Fine flowers
Brunfelsia pauciflora (lady-of-the-night)	6'	30 to 40	Flowers most of the year
Callistemon citrinus (bottlebrush)	25'	20 to 30	Lovely flowers
Cassia alata (candlestick plant)	9'	30 to 40	Lovely flowers
Codiaem variegatum (croton)	9'	30 to 40	Many varieties
Dombeya wallichii	10'	30 to 40	Fine pink flowers

Botanical & Common Name	Mature Height	Min. Night Temperature, °F	Remarks
Erica mediterranea (heath)	5'	20 to 30	Evergreen shrub
Euphorbia pulcherima (poinsettia)	12'	30 to 40	A good accent plant
Erythrina herbacea	5'	20 to 30	Deciduous shrub with showy flowers
Feijoa sellowiana (pineapple guava)	15–20'	30 to 40	Lovely flowers
Gardenia jasminoides (cape jasmine)	6'	30 to 40	Fragrance galore
Hibiscus rosa sinensis	10'	-10 to -5	Numerous bloom colors
Hibiscus syriacus	15'	-10 to -5	Favorite Florida shrub, many varieties
Ixora coccinea (flame-of-the-woods)	10'	20 to 30	Fine red flowers
I. 'Maui'	8'	30 to 40	Robust variety
Jatropha podogrica	10'	20 to 30	Excellent shrub
J. integerrima	10'	20 to 30	Fine accent
Lagerstroemia indica (crepe myrtle)	15'	20 to 30	Can grow into tree
Ligustrum japonica (Japanese privet)	20'	5 to 10	Evergreen shrub
L. sinensis (Japanese privet)	12'	5 to 10	Handsome shrub
Lantana camara	12'	20 to 30	Colorful flowers
Nandina domestica (heavenly bamboo)	8'	5 to 10	Fresh and green in droughts
Nereum oleander (oleander)	15'	10 to 20	Tolerates hot, dry conditions
Pentas lanceolata	8'	30 to 40	Vivid colors
Petrea volubilis (Queen's wreath)	15'	10 to 20	Fine blue flowers
Phyllostachys (bamboo)	20–50'	5 to 10	Handsome but invasive

The blooms of *Pentas lanceolata*. (*Barbara J. Coxe*)

Pentas—now available in several colors—are a fine shrub for Florida gardens and are used extensively. Varietal names vary. (*Barbara J. Coxe*)

The stunning bird of paradise. (*Barbara J. Coxe*)

The omnipresent crotons used so extensively in Florida come in a wide range of colors, each with their own varietal name; all are exotic foliage plants for Florida landscapes. (*J. Kramer*)

Not often seen in Florida—but certainly desirable—is the old-fashioned *Lantana camara*, a floriferous plant with umbels of orange flowers. (*J. Kramer*)

An hibiscus trained as a tree.
(*Barbara J. Coxe*)

A doubled form of hibiscus in a luscious butterscotch color.
(*Barbara J. Coxe*)

Yellow hibiscus are dramatic in the landscape and can be used as a hedge. Plants thrive in warmth and sun, and flowers are profuse.
(*J. Kramer*)

Jatropha species have handsome foliage and red flowers; there are several varieties. These are fine southern-climate plants that add much color to a landscape.
(*J. Kramer*)

It is tough to beat gardenias for beauty and fragrance (in any climate). These handsome shrubs can grow quite tall in the South. This is the standard gardenia, but there are also dwarf varieties.
(*J. Kramer*)

Botanical & Common Name	Mature Height	Min. Night Temperature, °F	Remarks
Plumbago capensis	10'	20 to 30	Blue flowers
Pyracantha coccinea (firethorn)	8'	5 to 10	Red berries
Raphiolepis indica	8'	20 to 30	Good for hedge
Russelia equisetiformis	8'	30 to 40	Colorful
Skimmia japonica (Japanese skimmia)	4'	5 to 10	Good in shade
Strelitzia reginae	8'	30 to 40	Famous bird of paradise
S. nicolai (white bird)	15'	30 to 40	Grows tall, lush
Tibouchina semidecandra (glory bush)	10'	20 to 30	Lovely violet flowers

Favorite Shrubs

Abelia floribunda has graceful arching branches with handsome bell-shaped pinkish flowers. It is quite pretty as an ornamental plant. This shrub is best grown in the northern part of the state, in zone 9. Smaller *Abelias*, growing to only 3 feet, are also available.

Abutilons are overlooked shrubs but display many pretty bell-shaped red or orange flowers. I grew them as container plants in California, but in Florida they do well as in-ground plants and add excellent color to the garden. These shrubs are somewhat spindly and grow to about 8 feet. They like sun and good drainage. Grow in zones 9 and 10.

The candlestick plant, *Cassia alata*, bears upright stalks of attractive yellow flowers. This low-growing shrub rarely gets taller than 8 or 9 feet. Its unique yellow "candlesticks" make a nice accent in the garden; grow in zones 9 and 10.

Brunfelsia pauciflora, or lady-of-the-night, is another good shrub not usually seen but very desirable in the tropical landscape. It bears violet flowers in profusion and is good as background accent. This spreading shrub is a bit sensitive to salty soil. It grows to about 6 feet and does well in zones 9 to 11.

Codiaeum variegatum, or crotons, are rampant in southern Florida, where they are used as hedges and accents. The plants have wonderfully colorful lanceolate foliage. The dozens of varieties love sun and water and need just moderate care. They grow well in zones 9 to 11.

The hibiscus is often called the "Hawaiian flower," but it could be called Florida's favorite flower as well. This genus contains about 200 species and numerous hybrids. The *H. rosa sinensis* family is the most popular in Florida, and the large flowers are available in an array of colors. *H.* 'Blue Bayou' is dramatic—a deep violet hibiscus that cools any landscape and looks luscious against deep greens. *H.* 'Norman Lee' is bright orange, *H.* 'Brilliant' is the red form we commonly see, and there are dozens of other varieties to choose from.

The plants are generally erect shrubs and are of manageable height, usually about 6 feet tall. Flowers are numerous, and good sunlight is essential for blossoming. Hibiscus like warmth—temperatures below 45 degrees F will injure most varieties. Fertilize hibiscus with a 20–20–20 plant food twice a month. Although plants can tolerate drought, even moisture is best for good growth. A stellar plant for southern Florida, the hibiscus is dependable and colorful in the landscape and is rarely bothered by insects when cultivated properly.

I like the flame-of-the-woods, *Ixora coccinea*, and have grown it off and on in many climates, both as a houseplant or in the garden. These amazing shrubs are content to grow in the shade and bear wonderful bouquets of the vivid red flowers that give them their common name. They like a well-drained soil and some pruning (they can grow somewhat rampant). Grow in zones 10 and 11.

If you really want color, try the many pentas, which bear magenta, pink, red, or white florets. The plants grow to 8 feet and are somewhat dense. Many varieties do best when grown in the sun, and most will survive heat well. Protect plants from cold. Grow in zones 9 to 11.

In northern California I was always trying to grow gardenias, but I was not very successful. In Florida, as garden plants gardenias offer fragrance and color and grow quickly to 8 feet in well-drained soil. The only problem with these shrubs is that so many varieties are available that it is hard to make a selection! Gardenias also come in all sizes, from midget to dwarf to standard size. G. *jasminoides* (Cape jasmine) and G. *j. veitchii* are the most popular species, growing to about 6 feet. G. *jasminoides* 'Radicans' is the miniature gardenia. Gardenias grow in zones 9 to 11.

The pineapple guava, *Feijoa sellowiana*, graces many a garden.

Generally a tree, but sometimes classified as a shrub, I have seen specimens as tall as 20 feet. Not often seen, this offbeat plant is worth trying in zones 10 and 11.

Tibouchina semidecandra, the glory bush, has dramatic dark violet flowers. It is a sprawling, not very attractive shrub, but its large flowers add a nice touch of purple to the landscape. It grows well in the shade and needs good drainage. This is strictly an accent plant that is fine for zones 10 and 11.

Although not as flashy as its cousin *Strelitzia reginae*, the popular bird of paradise, I prefer *S. nicolai*, which is almost indestructible and blooms with fine white-and-blue bird-like flowers. This plant grows well with even moisture and moderate light. This beautiful plant in my garden room still grows lushly after seven years of doing little other than top dressing the soil. With it's handsome flowers and spoon-shaped leaves, its only drawback is its rapid growth, reaching up to 15 feet.

Rhapiolpis indica is not widely known, but grows so well in southern climates that I recommend it for accents or hedge plantings. The plant's compact, bushy habit is perfect for creating hedge from several plants. The flowers, although small, are pretty. This shrub appreciates regular feeding and trimming and does well in a sunny location.

Eleven

BULBS

Bulbs are exquisitely beautiful and should not be overlooked when you are selecting plants for your Florida exotics garden. They are excellent subjects for any spot where you want color. However, do not group just a few in one area: grow masses for extravagant color.

Superb bulbs for Florida include amaryllis, eucharis, cannas, and many others. Most tropical and semitropical bulbs can be carried over each season, and the plants grow most of the year, resting for only a little while before blooming again. The secret is to start bulbs into growth slowly.

Most exotic bulbs do best in semishade and need a rich, humusy soil that drains readily. When planting, leave the tip (the narrow end) above soil, or cover the bulb halfway. Generally (but not always), the soil should reach the collar of the bulb; if it is above this area, rot may develop. After bulbs have been planted, water moderately the first few weeks, until some growth appears. Increase watering as leaves appear. Once established, most flowering bulbs will tolerate moderate drought conditions.

Feed plant every two weeks with a 10–10–5 food, and add a pinch of bonemeal to the soil to ensure good bloom. Should frost occur in your area, cover plants with burlap or place mulch around them.

Exotic bulbs are sold at nurseries at seasonal times or are available through mail order.

Favorite Bulbs

Amaryllis produce red, white, pink, rose, or striped flowers up to 7 inches in diameter. Let the upper third of the bulb protrude above the soil line when planting. Give plants little water until the leaves are about 1 foot tall, at which point start watering heavily. Amaryllis bear their lovely flowers in about one month. After plants flower, leave bulbs to grow on; they will multiply on their own. My favorite amaryllis are 'Appleblossom' and 'Red Fire', but there are more than 140 varieties.

Cannas have a lush tropical ambiance and bear large leaves with tall spikes of very colorful flowers. They bloom from early spring to fall in colors of orange, red, pink, and yellow. Sometimes called canna lilies, they are not lily family members. The plants do well in heat and love sun, copious water, and some feeding. The rhizomes can be left in the ground in warm climates.

Mail order catalogs list dozens of cannas, but here are some I have tried in Florida climates:

- *Canna* 'Chinese Coral' grows to 3 feet with fine coral flowers.
- C. 'Cleopatra' has variegated foliage and yellow flowers with red veining.
- C. 'Los Angeles' is a delicate pink flowering type.
- C. 'Madame Butterfly' brings yellow color to the garden.
- C. 'Panache', growing to 6 feet, has whitish flowers resembling an orchid.
- C. 'Previa' is a fine background plant with orange flowers and striped foliage.
- C. 'Red Stripe' and C. 'Red Ribbon' are red.
- C. 'Rosalinda' and C. 'Tiger Moth' bear yellow blooms.
- C. 'Rose Futurity' offers striking contrast with bronze foliage and scarlet red blooms.

Clivia produce bright orange flowers in the spring. When not in bloom, the dark green clumps of foliage are attractive accents in the garden. Use a rich soil that drains well, and grow plants in a semi-shady location. Water clivias evenly year-round. The bulbs can stay in the ground because the plants generally are not bothered by weather below 40 degrees F. The recommended variety is C. *miniata*.

Crinum make large plants with broad green leaves and wine-red or whitish-pink flowers. Plant several bulbs together to form clumps; leave half an inch of the bulb above the soil. Start plants in January or

Hymenocallis harrisiana is a frequent sight in Florida. It is a large plant with many leaves, and its flowers are spider-like with finely detailed petals. (*John Kilmer*)

The Nerine lily is frequently seen and grows well in warm climates. Many varieties are available. (*John Kilmer*)

Who can resist amaryllis? These plants with enormous flowers are not often seen in Florida gardens, but they deserve more notice. There are many varieties available; this one is 'Appleblossom'. (*John Kilmer*)

Certainly *Gloriosa rothschildiana*, with its scarlet-red flowers of incredible beauty, should be grown more. Classified as a vine, it is actually a fast-growing bulbous plant. (*John Kilmer*)

Crinums are handsome bulbous plants that are often seen in warm environments. Try using them en masse in landscapes where an accent is needed. (*John Kilmer*)

February; water them slightly until growth starts, and then increase watering. Crinums like a somewhat sunny spot. Keep plants in the ground year-round for flowers the following year. C. *moorei* or C. 'Ellen Bousanquet' are recommended varieties.

Tall wands of brilliant red flowers and grassy leaves make *Crocosmia* (or *montebretia*) a standout in the garden. Crocosmias need a well-draining soil in a sunny spot. They can stay in the ground year-round because they are generally frost hardy. C. *masonorum* is an especially good variety.

I heartily recommend the Amazon lily (*Eucharis grandiflora*) for Florida landscapes. The fragrant white flowers appear on foot-long stems about May or June. Plant bulbs about half their depth, in rich soil. Water little until growth starts, then increase the moisture. Grow Amazon lilies in a sunny spot, with good watering year-round.

Gloriosa rothschildiana, the glory lily, produces masses of red and yellow flowers on vining stems. Start tubes just below the surface, in sandy soil. Water moderately until leaves show, and then increase the watering. Provide support with a trellis or a post. Rest plants after they flower by watering them sparingly for a few weeks.

The brilliant red flower crowns of *haemanthus*, or blood lilies, are dazzling, but these bulbs are difficult to bring into bloom. Choose a shady location and plant the large bulbs with the top half of each bulb extending above the soil. Keep the soil evenly moist. H. *katherinae* blooms in the spring after foliage matures; H. *coccinea* grows all winter.

Hymenocallis is sometimes called the spider lily and is summer flowering with star-shaped leaves and fragrant white flowers. H. *harrisona* is the popular species, but many variants exist. Plant grow in large clumps and need grooming.

Lycoris have graceful spiderlike red flowers borne on tall stems. L. *radiata* is most often seen and makes a stunning landscape plant when massed in groups. Best in temperate zones. Bulbs can be started at almost any time and require good soil, plenty of water, and sun.

The small white or yellow flowers of *Ornithogalum*, which develop on leafless stems, make exquisite cut flowers. Grow Star-of-Bethlehem in semishade, in good-draining soil. When growth starts, increase watering until plants are receiving plenty of water. O. *arabicum* has white flowers with a black center; O. *aureum* has yellow flowers.

Scilla has bright green, grassy leaves and vividly blue spring flowers borne on short stalks. Grow scillas in shade, and be sure moisture is even. Leave bulbs in the ground year-round, but protect them from temperatures below 45 degrees F.

Sparaxis' grassy foliage and small, vivid red flowers with dark centers appear in the spring. Sparaxis like sun and even moisture at their roots. Bring plants into growth slowly, increasing water as leaves start growth. Grow masses of *S. tricolor*, the harlequin flower, as a wonderful accent.

Sprekelia, the Jacobean lily, has gorgeous red flowers in June just before its foliage develops. Plant bulbs in rich soil in a sunny spot, and keep the soil evenly moist. Water moderately most of the year, and feed plants every two weeks with a 10–20–10 food. *S. formosissima* is a good variety for most gardens.

From South Africa, *Vallotas* are evergreen plants that display beautiful red flowers after about one year. Grow plants in full sun in rich well-drained soil. Plant bulbs with their tips slightly above the soil line. Water moderately at first, then increase watering. *V. purpurata* is an excellent choice.

The pinkish, pendantlike flowers of *Veltheimias* appear in bunches. Start bulbs in January or February in sunny, well-drained soil. Water moderately at first; increase water as leaves open, maintaining even moisture at the roots. Feed the plants every two weeks. When flowers fade and leaves turn yellow let the plants die back naturally rather than cutting them off. Rest plants for two months. The variety I like best is *V. viridifolia*.

Twelve

VINES

Vines are considered a gardener's best friend, especially in Florida, where lovely *Bougainvillea* cascade over walls and brilliant *Allamandas* drench fences with color. These living screens of color impart a lovely old-fashioned look that softens the harsh lines of the garden and house. Vines create an interesting mass of color, cover unsightly areas, grow quickly, offer both foliage and flowers, have a cooling effect in a garden, and are easy to grow.

Most vines produce tendrils that grab on to any support, but some need to be tied to a support—a trellis, a wall, or a wire screen—with nylon thread, string, or vine clips. If vines grow too rampant, merely trim them back.

Buy vines from your local nursery. Do not plant vines that are extremely tender; grow those designated for zones 9, 10, and 11. March through September is the best time to plant vines. Most vines are tough and easy to get established. Vines need a rich soil that has been improved with humus and peatmoss. Dig planting holes deep—at least 3 feet to allow roots room to grow—and do not plant vines closer than 4 feet apart.

Water plants sparsely the first few weeks, and then water heavily, especially during the warm months. Vines growing in sunny areas will need more water than those growing in shade. Soil tends to dry out during the Florida summers because of reduced rainfall, so be sure to keep the soil moist.

Bougainvillea makes a stunning wall treatment. (*Barbara J. Coxe*)

Bougainvillea is the vine of choice in the South, and why not? It grows profusely and provides handsome color. This is *B.* 'Barbara Karst', but there are many varieties. (*J. Kramer*)

A lavender bougainvillea. (*Barbara J. Coxe*)

There are a great many passion flowers, and during the summer they bear their large, unique flowers. The red form bears a smaller flower than the usual type; its botanical name is *P. coccinea.* (*J. Kramer*)

Feed vines twice a month with a general plant food, such as 10–10–5. Because they often grow so quickly, prune vines back when necessary. I once put a small pot of *Thunbergia grandiflora* outdoors because it was not flourishing indoors. Within a year, lovely blue flowers were cascading over my fence, providing striking beauty and enjoyment. Trim vines as necessary, and train them to their supports. Most vines do not need constant pruning; check Table 12-1 for the best time to prune.

TABLE 12-1

Botanical & Common Names	Min. Night Temperature, °F	Remarks
Allamanda cathartica	30 to 40	Heavy stems, lovely flowers; prune in spring; sun
Antigonon leptopus (coral vine)	30 to 40	Fine screen; give light support: support; prune hard after bloom; sun
Aristolochia elegans (calico flowers)	30 to 40	Showy flowers
Bignonia capreolata (cross vine; trumpet vine)	-5 to 5	Orange flowers; thin out weak branches in spring; sun or shade
Bougainvillea	30 to 40	A favorite Florida vine
Clematis armandi (evergreen clematis)	5 to 10	Lovely flowers; needs support; prune lightly after bloom; sun
Clerodendrum thomsonae	30 to 40	Blooms over a long period
Doxantha unguis-cati	10 to 20	Yellow flowers; needs no support; prune hard after bloom; sun
Fatshedera lizei	20 to 30	Handsome foliage; no pruning needed; shade
Ficus pumila repens (creeping fig)	20 to 30	Heart-shaped leaves; thin plant in late fall or early spring; sun or shade
Gloriosa rothschildiana	30 to 40	Very showy and fast-growing
Gelsemium sempervirens (Carolina jessamine)	5 to 10	Fragrant yellow flowers; needs support; thin plant right after bloom; sun or partial shade
Hoya carnosa	10 to 20	Fine waxy flowers 6 months of the year

Botanical & Common Names	Min. Night Temperature, °F	Remarks
Jasminum nudiflorum	-10 to -5	Yellow flowers; needs strong support; thin and shape after bloom; sun or shade
J. officinale (white jasmine)	5 to 10	White flowers; needs strong strong support; thin and shape after bloom; sun or shade
Kadsura japonica (scarlet kadsura)	5 to 10	Red berries in fall; needs support; prune in spring; sun
Lonicera caprifolium (sweet honeysuckle)	-10 to -5	White or yellow flowers; prune in prune in fall or spring; sun
L. hildebrandiana (Burmese honeysuckle)	20 to 30	Shiny leaves; needs support; prune in late fall; sun or partial shade
Mandevilla suaveolens (Chilean jasmine)	20 to 30	Heart-shaped leaves and blooms; cut back lightly in fall; sun
Pandorea jasminoides	30 to 40	Very floriferous; fine white-pink flowers; grows to medium height
Passiflora caerulea (passion flower)	5 to 10	Spectacular flowers; needs support; prune hard; sun
Petrea volubilis	10 to 20	Fine blue flowers
Plumbago capensis (plumbago)	20 to 30	Blue flowers; prune lightly in spring; sun
Pyrostegia ignea (trumpet vine)	30 to 40	Showy flowers
Rosa spp. (rambler rose)	-10 to -5	Many varieties; needs support; prune out dead wood, shorten long shoots, and cut laterals back to two nodes in spring or early after bloom; sun
Stephanotis floribunda	20 to 30	Fragrant white flowers
Tecomaria capensis	30 to 40	Evergreen vine
Thunbergia grandiflora (blue trumpet vine)		Woody twiner with blue flowers
Trachelospermum jasminoides (star jasmine)	20 to 30	Small white flowers; needs heavy support; prune very lightly in fall; partial shade

Thunbergia grandiflora is a handsome but extremely fast-growing vine that can cover a wall or fence in quick order. The blue flowers, however, are spectacular and worth the effort of keeping Thunbergia in-bounds. (*J. Kramer*)

Pandorea jasminoides is a small-flowered vine that has delicate appeal. Flowers appear in clusters off and on throughout the summer months. (*J. Kramer*)

The bold orange of *Pyrostegia ignea*. (*Barbara J. Coxe*)

Allamanda verifolia appears in many southern gardens—sometimes as a hedge, other times as an accent. New varieties have small flowers and are less robust, but any Allamanda makes a fine addition to the garden. (*J. Kramer*)

Favorite Vines

I must admit that I dote on vines, to the point where I do not consider any garden complete unless it has some fabulous vines twining through the scene. Few plants can match the beauty of bougainvilleas in bloom, or the unequaled fragrance of jasmine.

Bougainvillea, the flower of the South, is available in many varieties, with red, white, orange, or pink flowers. To me, 'Barbara Karst', a single red, is still the star. Other exotic and desirable bougainvilleas are 'Raspberry Ice', a violet-red, and 'Scarlett O'Hara', a fiery red. These plants like sun, water, and feeding; they will grow quickly and cover unsightly areas while adding a tropical note of color. Most bougainvilleas need support, such as an arbor or a trellis. If you plant against a house wall, plan on not painting the wall for quite awhile, because bougainvilleas are hard to remove from brick or plaster surfaces.

Gelsemium rankinii, swamp jasmine or jessamine, is a vigorous climber that sports wonderful clusters of yellow bell-shaped flowers. This easy-to-grow vine is handsome but needs support. It is suitable for zones 9 to 11.

True jasmines (*Jasminum* sp.) perfume the air of Florida. Three stand-out species are *J. officinale*, with fragrant white flowers; *J. polyanthum*, with pink flowers; and *J. multipartitum*, also with pink flowers and suitable as a ground cover. Jasmines can tolerate some drought but prefer plenty of moisture, good soil, and light feeding. All three species will take moderate cold (40 degrees F), but not for too many nights, so be prepared to protect them if necessary. Most jasmines grow quickly and can become bushy and shrublike.

The passion flower bears large, exotic flowers; *Passiflora pfordtii* and *P. vitifolia* are the two species most often grown. Both have red flowers, but the dramatic flowers of *P. vitifolia* are especially attractive. Passion flowers like plenty of water, and may grow as high as 40 feet. The lobed foliage is quite handsome. These vines are fine for zones 9 to 11.

Stephanotis floribunda prefers cool nights (to 50 degrees F), so to thrive in Florida, this vine needs to be grown in a shady protected spot to encourage it to bear its spectacular waxy white, fragrant flowers. The plants have few leaves and love to climb, so give them proper support. Stephanotis takes a while to become established; it is fine for zones 9 to 11.

The orange cape honeysuckle is all over Florida, growing on fences and trellises. *Tecomaria capensis* is a vigorous grower and climbs quickly. Grow it somewhat dry in a sunny place for best results. Sometimes classified as a shrub, this honeysuckle is good for zones 10 and 11.

Thunbergia grandiflora is a tremendously fast grower; within weeks it can engulf anything it grows over, but its myriad blue flowers make it a welcome sight. It is fine for zones 10 and 11.

Pyrostegia ignea is a favorite vine in Florida because of its orange, tubular-shaped flowers. This robust plant will cover walls with sheets of color in January. Grow it in zones 9 to 11.

Thirteen

Bananas and Gingers

With their lush foliage and vibrant blooms, gingers and bananas personify the tropics. These dramatic exotic plants are finally taking their rightful place in Southern gardens, making ordinary gardens into Edens of delights.

Bananas

The bananas have never tempted me as an ornamental, but lately some unusual varieties have come to market, with attractive foliage and great variation in fruits. I must admit I am fond of the red flowering banana called *Musa coccinea* with its fine red flowers; it bears no fruit but the flowers are exceptional. The dwarf banana called M. *cavendish* and the newer variety, 'Apple', are quite handsome and do bear fruit. The tropical foliage is lush and attractive.

Bananas are native to tropical Africa, Australia, and the South Pacific, and there are about 30 species in the group. Bananas are not trees but rather perennial shrubs growing from underground corms or rhizomes. The fleshy stems are sheathed with broad green leaves that can grow to 20 feet and each stem produces one flower cluster which bears fruit and then expires. New stems keep growing from the rhizomes. Bananas need warmth, never below 55 degrees F at night. If bananas get caught in a freeze, they do not die altogether but will begin growth anew come spring and summer.

In addition to M. *coccinea* and M. *cavendish*, Stoke's Tropicals lists dozens of banana varieties, including Golden Yellow, Hua Moa, Kru, Orinoco, Blúggoe, Nino, Rajapuri, and on and on.

Gingers

This overlooked family is finally gaining popularity in gardens; they have wonderful color, symmetrical growth, and heavenly perfumes. Gingers are perennial herbs and offer great ornamental use only recently recognized by landscape people. They are from the diversified *Zingiberaceae* family and like plenty of warmth and lots of water and feeding during their growing time. Many generally go dormant in winter and can be left in the ground in warm climes. The plants are generally bought as rhizomes, although lately some are available as pot plants.

I have always been fond of the gingers (which include curcumas, alpinias, and hedychiums) and have grown them as pot plants in northern California. In Florida, I have them in my garden and lanai, where they perform so well.

Kahili gingers, called hedychiums, with their plumes of flowers signify tropics and bloom with abandon in warmth. Heliconias, which belong to the Zingiber family, offer plumes of orange flowers, and are sometimes called parrot flowers; en masse they are a terrific sight. The gingers called pine cone ginger (Zingiber species) have aromatic leaves and creamy petals; the most common species, Z. *zerumbet*, has red bracts and tiny white flowers.

Here are some of the best gingers to plant in Florida:

Alpinia (Shell Ginger)

The Alpinias are known for their delicate spikes of white, red, or pink flowers. They make good cut flowers. Easy to grow, these plants thrive in sun and need lots of water.

A. *Garanga* is the culinary spice cardamon and makes an excellent plant for both containers and in-ground use. The red ginger, A. *purpurata*, with showy flower spikes of cerise bracts is very pretty. A. *zerumbet* has pendant shell-like whitish pink flowers. Varieties of A. *zerumbet* with variegated foliage add drama to the landscape or in a container.

Shell ginger, *Alpinia zerumbet*, is known for its waxy flowers in pendant growth. It grows like a weed in warmth and sun. (*J. Kramer*)

Kahili ginger, *Hedychium gardnerianum*, brings golden yellow flowers to the garden and is a handsome plant that should be grown more often. (*J. Kramer*)

Red torch gingers, *Alpinia purpurata*, grow into large plants that bear handsome spikes of red flowers. They are used as cut flowers in many areas. (*Fred Berry*)

Costus comprise a large group of brilliantly colored exotic flowers; this unidentified species shows the tall spike formation and clustered cone. (*Fred Berry*)

Costus (Spiral Ginger)

The fantastic plants of the Costus family produce stunning flowers. The family consists of 4 genera and about 100 species distributed throughout Asia, Africa, and the tropics. The stems bear spirally arranged leaves of symmetrical beauty. The waxy flowers are brilliant bracts of red, much like bromeliads. These may be the prettiest of all gingers and like dappled sun and plenty of moisture with good drainage and some feeding. Costus plants are beginning to be available in nurseries as potted plants. They make fine landscape subjects and add allure to a tropical garden.

Curcuma

Curcumas are not well known but offer great flower color. They grow to 4 feet and have large showy cones of small flowers. They go dormant in fall and winter and like dappled sun and moderate waterings. I find curcumas especially fine, easy to grow plants and a recent bed of the plants at the Naples Beach Hotel in Naples, Florida evoked many letters of comment about their beauty. A definite must for lovers of the exotics.

C. *elata* has pale pink rose bracts on tall stems and is a colorful addition planted in mass. C. *petiolata* has wide leaved foliage and pink bracts; it grows quickly into a handsome plant. C. *roscoeana* has papery orange bracts and blooms abundantly in semi-sun. C. *undatus* bears a large waxy inflorescence and has very attractive foliage.

Hedychium (Butterfly Ginger)

Hedychiums have great fragrance and great beauty and bear plumes of showy flowers on top of tall canelike stems. These gingers are wonderful in dappled sunlight and like ample watering, but don't flood them. Feed moderately. You can find some of these varieties as plants at local Florida nurseries or order by mail.

H. *coronarium*, known as butterfly ginger, has been grown for years as a houseplant in the north. It has fragrant white flowers, and it is very pretty for the landscape. H. 'Fiesta' has deep yellow flowers with a red throat. A good bloomer throughout spring and summer, it comes highly recommended. H. *flavum* is the yellow butterfly ginger; it grows quite tall, up to 8 feet. H. 'Kahili', sometimes called H. *gardenerianum*, blooms bright yellow.

Kaempferia

Small kaempferia gingers are sometimes called peacock ginger or Asian crocus, and make fine garden subjects with both handsome flowers and foliage. They like morning sun, full water, and some feeding. When they die down, water sparingly, and protect against cold temperatures (below 45 degrees F).

K. decora is known as the dwarf ginger lily. It makes a fine display of bright canary yellow flowers. *K. masonii* has round leaves and purple flowers. *K. rotunda* is sometimes called the resurrection lily; it has fragrant flowers with lilac tips. It prefers dappled sunlight.

Zingiber (Pine Cone Ginger)

Zingibers are seldom seen in Florida gardens but offer good flower color, do very well in tropical climates, and make handsome landscape subjects. They need dappled sun and even moisture at the roots. Give them a winter rest after blooming.

A large family with many types, zingibers include *Z. spectabilis*, the pine cone ginger, with red bracts and dark green leaves; it grows to 6 feet. *Z. darcyi* is a cousin with variegated foliage and tiny white flowers. It attains a height of 5 feet.

Heliconias

Heliconias are in a family of their own—Heliconiaceae. Most have broad, lush leaves and colorful inflorescences that are so vibrant in color they seem artificial. Currently called parrot flowers because the flower bracts are so brashly colored, the heliconias are favorites of mine and I grew them as houseplants without much success in northern climates. Here in Florida they thrive in my garden, tall and stately and full of orange and red, or yellow and red, or brilliant red, green, and yellow flowers. There are many, many types and all deserve recognition. I have grown them in full sun with ample water and protection from cold, anything below 55 degrees F.

Heliconias are only now coming on the market as landscape plants. Among some of the best are *H. caribaea,* which grows to 15 feet and has red and white bracts appearing in the summer. There are dozens of varieties of this species, all colorful and brilliant additions to the garden. *H. bihai* grows tall to 15 feet. Its flower spikes are not as brilliant as *H. caribeae*, but the type known as 'Lobster Claw' is stunning.

H. stricta is the most commonly grown species; its bracts are waxy and usually red and yellow or green. I started growing *H. psittacorum* in Chicago in 1965. There are many varieties, all of which bloom with orange bracts resembling small birds of paradise. I was thoroughly

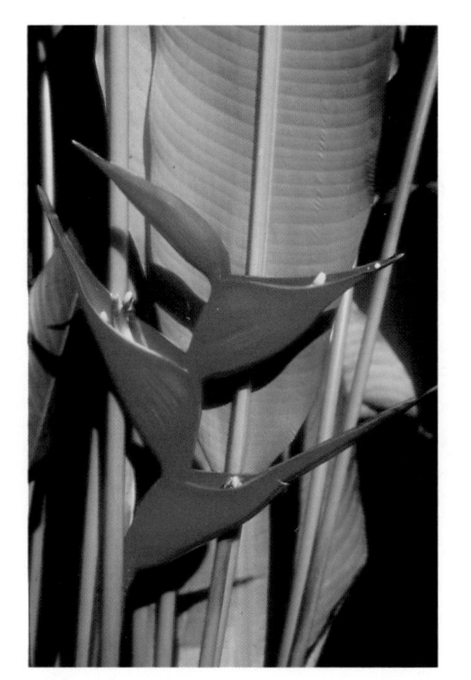

Heliconia stricta, from the banana family, offers striking parrot-like flowers in vibrant color. Used as garden accents, these exotic plants are always show-stoppers. (*Fred Berry*)

This *Heliconia psittacorum* hybrid offers blatant red and yellow color. It is a regal plant with tall spikes of flowers that is suitable to warm environments. (*Fred Berry*)

Musa velutina is the banana as we know it—it bears a unique flower spike and fruit formation. This perennial plant dies down and grows again come warm weather. (*Fred Berry*)

enchanted with them in Chicago and am more so now as they bloom in my Florida garden with zest.

H. cuminata bears white and red bracts, and I have seen this bloom frequently at Selby Gardens in Sarasota, Florida.

There are pendant flowering heliconias as well; space does not allow me to enumerate all of them. For best reading consult *Heliconias: An Identification Guide* by Fred Barry and W. John Kress published by Smithsonian—a wonderful compendium of heliconias by the experts.

Fourteen

WATER LILIES, LOTUS, AND AQUATICS

Every tropical garden pool needs exotic lilies and lotus. The most colorful and well-known of all genera in the water lily family are *Nymphaea*. Extensive hybridization has produced flowers of magnificent shape and color in hundreds of varieties, each more beautiful than the others.

Nymphaea are divided into two classes: the hardy lilies with medium- to large-sized flowers, and the tropicals, with larger and more colorful flowers. Almost all tropical lilies are fragrant; some open by day and others by night. In hardy lilies, colors range from white to pink, yellow to red; in the tropicals we find the breathtaking blue and purple hues.

Some outstanding hardy *Nymphaea* hybrids were raised by M. Bory Marliac towards the end of the last century. In the tropical lilies, Mr. George Pring of the Missouri Botanic Gardens has raised some superlative varieties.

Once planted in rich soil in a sunny location, water lilies require little care. The tropicals are somewhat more difficult than hardy varieties to get started, but once growth is under way, they need little care.

The foliage of water lilies grows continuously through summer, and feedings in June and July will produce heavy growth and abundant flowers.

Planting Tropical Water Lilies

These fabulous tender *Nymphaea* put on a special show of breathtaking color. The crisp, cool, green leaves and exquisite flowers are a magnificent addition to any pond or pool.

Tropicals can be set out as soon as the water in their pond is warm, about 72 degrees F. Container culture is recommended for the tropicals. These lilies are sold as growing plants, and the tubers are wrapped in earth or moss. Do not break the earth ball; plant it whole in soil in a container or in the bottom of a pool. Hold the plants in place with a stone, and cover the soil with gravel to keep the soil in place under the water. Be sure the roots are placed horizontally in the container, about 1 inch below the surface of the soil with the growing tip peeking from the soil. If possible, place tropical lilies so that 4 to 6 inches of water covers the crowns in the first month. Then lower them so that there is 8 to 10 inches of water above the crowns.

Planting Lotus

Lotus (*Nelumbo* spp.) command interest and admiration with their spectacularly large flowers on tall stems. A native of India, the lotus was sacred to the Hindus, who considered it the most exalted flower. The plant also appears as a basic motif in Persian, Assyrian, and Egyptian art.

The lovely, silvery blue-green leaves are often 30 inches in diameter, and plants bloom from the end of June to the end of August. The foliage (like shallow bowls) does not have the customary notch at the junction of stem and leaf. The flowers sometimes reach 12 inches in diameter and are quite fragrant. They open just slightly the first day, close, then open again the following morning, and remain open for several hours before they close again. On the third day they open fully.

Plants grow quickly, and good-sized roots may bloom in the first season; satisfactory flowering occurs in the second season. Lotus are hardy as long as the roots are below the frost line (about 18 inches of water). In containers, they will be perfectly safe through winter, provided ice does not touch the crown of the plant.

Plant lotus in the spring when warm weather has started. Be careful not to break the growing tip of the tuber. For best growth put plants in a wooden tub, box, or half barrel. Use large containers at least 24 inches in diameter; if roots strike the corner of a planter they may die.

Nymphaea 'Emily Grant'. (*Dency Kane*)

Water lilies and lotus decorate this water scene and provide colorful beauty. (*A. R. Addkison*)

To plant lotus, make a hole in a container filled to within 4 inches of the rim with soil and fertilizer. Place one root horizontally in the trench, and cover with 2 inches of soil; leave about a half-inch of the growing tip exposed. Put a flat stone on the covered root, and add more sand, being careful not to touch the growing tip. Put 1/2 pound of fertilizer in the bottom of the tub or pool. Be sure the lotus is covered with 3 to 4 inches of water, and repot plants every other year in a fresh supply of soil and fertilizer. Without feeding lotus will not grow.

Lotus stock is generally limited at most nurseries, so order before May.

Botanical classes of lotuses (*Nelumbo*) change frequently, and there are several disagreements among authorities about names. For a simple workable class we list plants as follows:

- *Nelumbo lutea* (American lotus). Lovely light yellow flowers.
- *N. lutea* var. *flavescens*. Smaller blooms with red spots at base of petals.
- *N. nucifera* (Asiatic lotus). Handsome pink flowers with many varieties:
  var. *rosea plenum*: Large, pink, double blooms.
  var. *pekinensis rubra* (rosea): Rosy-red flowers.
  var. *pygmaea alba*: Dwarf, with white blooms.
- *N. Shiroman*. A beautiful white double lotus; huge flowers; free blooming.
- *N. Kinshiren*. A white-flowering plant with touches of rose color.

Plant Protection

You will find that water lilies and lotus are amazingly free from insect attacks and disease problems. Planted in sufficient water and full sunlight, they are vigorous, healthy plants, rarely bothered with the problems that their earth-borne relatives may suffer. However, a few insects and fungi occasionally attack plants, namely aphids, which are round-bodied pests that are easily seen. They appear on leaf stems and eventually move on to the leaves and suck vital plant juices. They are easy to eliminate by washing them off with a strong spray from the garden hose.

Caddis flies feed on the roots, buds, and leaves of just about anything that grows in water. The best control is to have goldfish in the pool; caddis fly larva is their favorite food.

Leaf rollers are larval pests that eat leaf edges. They can cause considerable damage to plants unless caught early. Control is simple: Pick off infested leaves or cut away rolled-up leaf edges.

The small leaf beetle hibernates in poolside vegetation. In June, it lays ovate eggs on leaf surfaces. The larva feed on leaves, and their work can soon be seen. Goldfish are the best control; they eat the larva and eggs.

Fungus diseases, although very rare in the small garden pool, occasionally attack water lilies and lotus and cause dark patches on the leaves. The best remedy is to remove the affected foliage immediately and burn it. Another fungus attacks stems, causing a blackening and rotting away of the growth. Copper sulfate is the usual remedy, but use it with utmost caution—a half-ounce placed in a small cloth bag for a pool that holds 2000 gallons of water. Pull the bag around the pool until the crystals have dissolved.

Aquatics and Marginal Plants

When we plan a water garden, we immediately think of water lilies. However, there are many other beautiful plants—iris and grasses for the water's edge, and floating aquatics, such as water hawthorn and duckweed. Ferns with lush green fronds are attractive, too, and especially suitable near water; their graceful growth and emerald colors always command attention. They require little care and thrive on the shady margins of the pool.

Native orchids and some of the pitcher-plants are unusual and tempting. Beautiful though they may be, I have tried them several times but have never been successful with them. But my failures may be your successes, so if you are fond of these lovely flowers, by all means, try some. (A list follows later in this chapter.)

Any pool is enhanced by the addition of some plants. Just what you decide to grow depends on your individual site and the kind of pool you have. The informal pool must have some greenery to make it appear natural, and the formal pool, too, can be decorated with a few container plants. Bright foliage and colorful flowers are an integral part of the water picture.

Marsh and bog plants include some beautiful species, and they will thrive in moist poolside conditions. The plants cover the demarcation between the pool and the soil and at the same time provide a handsome setting for water lilies.

Border plants will grow in any good heavy loam and need little attention. Putting them in place at their proper planting level is important. Some like their root crowns just slightly above water level; others must have their roots in marshy soil. Several need 6 to 10 inches of water above the crown of the plant.

Although these are called marginal plants, most of them can be used in planting boxes, too, where they are especially handsome as bright spots of color and to provide vertical accent.

Here are some of the more common plants for marginal plantings around the pool:

- Arrowhead (*Sagittaria*) is perhaps the most beautiful poolside plant. The dark green leaves are bold, and the white flowers have dainty yellow centers. Plants grow up to 4 feet in height and are spread by runners. Thin them occasionally if they get rampant. They grow best in wet soil or in water to 6 inches in depth. Common arrowhead (*S. latifolia*), giant arrowhead (*S. sagittifolia*), and Japanese arrowhead (*S. japonica florepleno*) are all highly recommended.
- Cat-tail (*Typha latifolia*) is a common bog plant that grows to 6 feet, with dark brown tails. It needs about 6 inches of water and reproduces by creeping root stock. Put it in a container to confine growth. *T. latifolia* is the common cat-tail; graceful cat-tail *T. angustifolia* is shorter, with narrow leaves and graceful tails; and pygmy cat-tail *T. minima* only grows to 12 inches.
- Egyptian paper plant (*Cyperus papyrus*) is a graceful member of the sedge family. It bears fronds of thready umbrella-shaped leaflets at the ends of bending stems. It can grow to 10 feet. *C. alternifolius* is also desirable.
- Horsetail (*Equisetum hyemale*) is one of the most popular pool plants and really a beauty with its tall segmented leaves of apple green. A lovely vertical accent, it can be controlled in a container. Horsetail likes about 1 to 4 inches of water.
- Taro (*Colocasia esculenta*), or elephant ears, is a tuberous herb that can reach to 3 feet, with magnificent foliage shaped like elephant's ears. Plant it with the root crown at water level. It needs a shady place and is a capable performer for poolside. Green taro

This reflecting pool with water lilies is the focus of the garden plan; in the rear is a bench on which to sit and enjoy the view. (*A. R. Addkison*)

Once planted, water lilies require little care and add much beauty to a landscape. (*A. R. Addkison*)

Water lotus blooming in a pond. (*A. R. Addkison*)

(*C. indica*) grows to 2 feet; imperial taro (*C. antiquorum illustris*) has splotched foliage to 4 feet; violet-stemmed taro is *C. violacea*.

- Water arum (*Calla palustris*) grows to 12 inches, with roots in mud slightly above or below water level. It takes a while to become established and produces white flowers in its second year. Not outstanding but desirable.
- Water canna (*Thalia*) is a perennial herb similar to canna but with bold, spear-shaped foliage and deep purple flowers on long arching stems. It flourishes in an inch or so of water. *T. dealbata* grows to 4 feet; *T. divaricata* is a larger form.
- Sweet flag (*Acorus calamus*). A hardy marsh perennial that likes shallow water, sweet flag has broad, dark green leaves to 3 feet. A small greenish flower is borne on a tall spike, making this a charming decoration.
- Iris. Many beardless irises grow well in heavy, moist soil. The appealing flowers start early in May and are bright accents near the water. *I. versicolor* is the popular and exquisite blue species; *I. kaempferi*, the Japanese iris, bears brilliant purple flowers and is not to be missed. *I. laevigata* is also handsome and similar to *I. kaempferi*. *I. siberica* is tall, to 4 feet, and also has blue-purple blooms. All need some sun.
- *Gunnera chilensis*. A very large waterside foliage plant with exquisite leaves. Needs rich and loamy wet soil. Crowns of plants must be above water and dry in the winter. Not hardy in the north. Very similar and equally handsome is *G. manicata*.

Bucket and Tub Gardens

For gardeners who admire and want water lilies but do not want to go through the work involved in the construction of a formal or informal custom pool, the tub garden is an ideal container. Containers used as miniature water lily pools are usually made of tough fiberglass and will last several years. The maximum size is 24 inches at the top tapering to 20 inches at the bottom and 20 inches deep. A container this size will hold one or two lilies. Set the container on the ground and fill it with 8 to 10 inches of soil, then plant. The ideal depth of water over the crowns of most water lilies and lotus is from 8 to 10 inches.

Pygmy lilies (*Nymphaea tetragona*) are especially suitable for the tub garden, especially 'Gloriosa', 'White Laydekeri', 'Aurora', 'Joanne Pring', and 'Carmine Laydekeri'. If you want to try tropical lilies, plant 'Blue Pygmy', 'Royal Purple', or 'Zanzibar Blue'.

Other plants for a tub garden are water poppy (*Hydrocleys nymphoides*) and water snowflake (*Nymphoides indica*). They do best planted in water from 5 to 6 inches deep and in 6 inches of soil or more.

Fifteen

BROMELIADS

Bromeliads are perfect in Florida for year-round color in the landscape. Aechmeas, billbergias, guzmanias, neoregelias, vrieseas, and tillandsias are the most popular bromeliads. Most bromeliads are from South America, although some are native to the United States. Some of these pretty plants—namely the aechmeas, billbergias, and vrieseas—have a vase shape that holds water and acts as a reservoir in case water is in short supply.

Most bromeliads are medium-sized plants, growing to about 30 inches. The true flowers are insignificant, but the bracts display brilliant colors. The bracts of the billbergias are cerise or apple-green; vrieseas bear erect stalks of orange plumes; and the neoregelias have bracts that turn red at bloom time.

In many bromeliads, equally as important as the bracts is the foliage, which is unusual and vivid, in subtle shades of gray-green, yellow, or mahogany. Some species have mottled foliage, others horizontal or vertical stripes.

Bromeliads need good air circulation to thrive. Plant bromeliads in equal parts of fine grade fir bark (sold at nurseries) and soil, in wide holes that are about 12 to 14 inches deep. You can also grow some plants above ground level on a bed of sphagnum tied to trees (use nylon cord to anchor the plants). Bromeliads need only water and sun;

feeding is generally not necessary. Here are my recommendations for Florida growing:

Aechmea

Requiring almost no care, these small, medium, or large epiphytic plants are usually vase-shaped, with glossy, brilliantly colored, variegated leaves. The flower spike is usually long, with small flowers hidden in the bracts. Many aechmeas bear white, red, or blue berries that last for several months. Give them bright light and pot them in fir bark or potting soil. Do not fertilize them. Keep the vase formed by the leaves filled with water. Flush out water and replenish once a week. Most species bloom in spring or summer, a few in winter. When flowers fade, suckers appear at the base of the plants. Cut these shoots off when they are 2 to 4 inches high, and pot them separately to propagate new plants.

Some of the aechmeas include A. *calyculata*, which grows to 20 inches and has flower heads of vivid yellow in mid-spring. A *chantini* grows to 36 inches, producing a large flower head and red bracts.

A. *fasciata* grows to 24 inches, with a tufted blue-and-pink flower head in spring. A. 'Maginali' is an outstanding hybrid with pendant red flowers, usually in winter, followed by blue-black berries; it grows to 30 inches. A. *pubescens* produces a wheat colored flower head in the fall, followed by white berries. It is a smaller variety, with a maximum height of 20 inches. Another small variety, A. *racineae*, grows only to 14 inches; it bears red, black, and yellow flowers at Christmastime.

A. *ramosa* can reach 40 inches in height; it produces a pyramidal head of yellow flowers with red bracts in summer. A. *weibachi* grows to 20 inches and has lavender flowers with red bracts, usually in winter.

Ananas

Ananas is the genus of the pineapple, namely A. *cosmosus* and A. *bracteatus*. The most popular ornamental ananas is probably A. *nana*, a novelty plant with rosette growth that does produce a tiny pineapple. These plants are not particularly handsome, but they do create a curiosity in the landscape. They grow well with sun and warmth.

Billbergia

Decorative plants with gray-green, silver-green, or purple foliage and bizarre bracts, billbergias are perfect for indoor growing and warm land-

Billbergia vittata 'Rubra' bears handsome flowers bracts at bloom-time and is frequently found in warm, sunny climates as a landscape plant or a lath house subject. (*J. Kramer*)

Small pineapples can be seen on this bromeliad botanically called *Ananas cosmosus* (this is the variegated species). These plants have thorny leaves and are troublesome to handle, but are still popular. For easier handling, ask for *A. nana*, a smaller version that bears tiny pineapples. (*J. Kramer*)

Ananas cosmosus variegata, one of the pineapple bromeliads, with its fountain growth and striped leaves. (*J. Kramer*)

Aechmea chantini, with its blatantly colored bracts and large colorful blossom head, is the queen of the Bromeliads. (*J. Kramer*)

Aechmea fasciata is perhaps the most widely grown bromeliad. It has a wonderful crown of pink bracts concealing tiny hidden blue flowers, and its leaves are banded and frosted. Many new varieties have appeared in the last few years. (*J. Kramer*)

scapes. The true flowers are small, but the colorful bracts are striking—red, or pink, or purple. Pot them in fir bark or an osmunda-and-soil mix. Give them bright light and keep the vase formed by leaves filled with water. They are good for planters in public places. Propagate them from offsets.

Among my favorites are *B. amoena*, which grows with shiny green leaves and rose-colored bracts in spring and summer. *B. brasiliense* grows to 48 inches. It has tubular gray-green leaves and handsome pendant cerise flowers; highly desirable. *B. morelii* is only 10 inches tall, with green leaves, blue flowers, and red bracts in summer.

B. nutans can reach 30 inches; it produces chartreuse, pink, and cerise flowers in winter and is known as "queen's tears." *B. pyramidalis*, which grows to 24 inches, has golden-green leaves, orange-pink flowers, and bracts in summer. A named cultivar *B.* 'Fantasia' is a robust hybrid with multicolored leaves, red bracts, and red-and-blue flowers in fall.

B. zebrina grows to 40 inches with lovely gray-green leaves flecked with silver and cascading rose-colored bracts, usually in summer.

Guzmania

These small- to medium-sized bromeliads almost outrank aechmeas as the best bromeliads, and rightly so because there is an incredible variation of leaf color and pattern available within the *Guzmania* genus. The decorative leaved beauties, such as *G. lindenii* provide startling room accents, and even seedling plants are colorful.

With about 135 species, guzmanias grow in the forests of Ecuador and Columbia; other species come from Central America, Brazil, Costa Rica, and Panama. The plants grow in varying elevations, and it is not unusual to find guzmanias at over 8,000 feet where nights get very cold. Generally, guzmanias grow best in shady situations atop tree limbs protected from direct sun—they grow on lower limbs rather than treetops. Being mainly epiphytic, the plants resent soil at the roots, although a few do grow as terrestrials in nature. Plants have smooth-edged, glossy green leaves, solid or with stripes, bands, spots, or crossbands in varying colors—indeed, there is a tapestry of color here. The basic growth pattern is a rosette, and most plants are about 3 inches across, making them fine for indoor spaces. The flower spike is erect and bears brilliantly colored bracts, from yellow to flaming red and burnt orange. The bracts last for some time. The tiny flowers are white or yellow.

My guzmania collection at an east window grows very well and produces beautiful color. They are in clay pots in fir bark. I flood the plants almost daily in very hot weather but do not water much in cooler weather. In any kind of weather I always keep the vases formed by the leaves filled with water. Do not feed these plants; those guzmanias that I fed reacted poorly and leaf burn can result. I mist plants frequently with tepid water. Pests rarely bother these bromeliads.

Unlike most bromeliads, guzmanias are not free with offshoots, producing only a few offshoots, or *kikis*, and are generally slow growing compared to the billbergias. There are small plants, medium growers, and a few large ones to 48 inches across. Here is an incredible array of good plants for all to try.

- G. *berteroniana* is a 24-inch rosette of shiny green leaves. The showy inflorescence has orange-red bracts with yellow flowers.
- G. *lindenii*, a 40-inch rosette, has spectacular green foliage marked with transverse wavy lines of dark green, red beneath. The erect flower scape has green bracts and white flowers. Likes coolness.
- G. *lingulata* is a handsome species about 18 inches across with a star-shaped, orange flower head.
- G. *l. major*, growing to 30 inches across, has a larger flower.
- G. *l. minor*, to 24 inches across, has a red-orange inflorescence with white flowers.
- G. 'Magnifica' (*lingulata* x *minor*) is a hybrid, 30 inches, green leaves with a slightly raised, fiery red flower crown with yellow flowers.
- G. *monostachia*, about 24 inches across, has satiny green leaves arranged in a dense rosette. The pokerlike flower spike is erect with white flowers and green bracts stenciled with maroon lines. The very tip of the inflorescence is crowned blood-red in many varieties, orange in others. A very showy bromeliad.
- G. *musaica* is sure to please houseplant enthusiasts. The leaves are 24 inches long, bright green, banded and overlaid with irregular lines of dark green and wavy purple markings on the reverse. The flower spike is erect and turns red at flowering time; white waxy flowers are set tight into the poker-shaped flower head.
- G. 'Orangeade', typical of the G. *lingulata* group, has a loose 30-inch rosette of green leaves and an exquisite, brilliant-red flower crown.

red-and-purple inflorescence. *H. stellata* has 3- to 5-foot spiny leaves and a tall spike with red bracts and purple flowers. An outstanding sight in any garden.

Neoregelia

In Germany and Japan, neoregelias are favorites indoors because they are handsome, undemanding, and among the finest foliage plants available. There are about 60 known neoregelias, most native to eastern Brazil; however, Columbia and Peru also have a share of the plants. In nature the plants grow near the ground or in lower branches on trees, preferring a shady place with good air circulation.

Most plants are medium-sized and grow in a compact, flat rosette. Some are only a few inches across; others grow to 5 feet. The brilliant foliage is plain green, spotted, marbled, striped, or banded with color, and at bloom time most plants show bright red to rose at the center. The flowers are small, usually in shades of blue or a combination of blue and white. Flowers die quickly, but the flush of color in the foliage remains for many months. Neoregelias are fine for northern or western exposures, where they get enough light but not too much direct sun. I purposely keep them at ground level so their beautiful foliage can be easily seen.

Use a potting mix of one-third perlite, one-third soil, and one-third fine-grade fir bark that drains readily. Flood plants with water during the warm months, but they are given less moisture in cool weather. I never feed neoregelias, but I always keep the vase filled with water. In my plant room small insects and frogs live in the center of the plants—those insects that die in the plant are excellent nutrition for the frogs and the plant. Unfortunately, the water reservoir can attract mosquitoes, so flush vases periodically. Occasionally wipe leaves with a damp cloth and spray foliage with tepid water, especially in the hot weather.

Neoregelias are rarely bothered by insects and are free with their offshoots, as many as four or five *kikis* per plant. Some of the best are listed here.

- *N. ampullacea*, about 9 inches across, has leaves with mahogany crossbands and small blue flowers in spring or summer. Give a little more light than for most neoregelias.
- *N.* 'Bonfire' grows to 20 inches in diameter with a beautiful

rosette of reddish-plum leaves. Flower crown deep in plant; tiny violet petals.

- N. *carolinae* is perhaps the showiest in the genus, with tapered leaves dark green and a rosette about 30 inches across. The center of the plant turns red before blooming. My plant was in full color for nine months.
- N. c. 'Meyendorfii' is a broad 30-inch rosette of flat olive-green leaves with coppery tinting. At flowering time the inner leaves turn a dark maroon; flowers are lilac and deep in the center.
- N. c. var. *tricolor* is expensive, but a panorama of color; the variegated leaves are white-striped. When in flower the foliage has a pinkish hue; the heart of the plant turns cerise. A highly desirable 30-inch bromeliad that steals the show. Grow it in shade with warmth.
- N. *compacta*, is a dense, erect rosette to 24 inches of green leaves; inner leaves turn red at bloom time. Good small plant.
- N. *concentrica*, about 30 inches across, has pale-green leaves flecked with purple and edged with black spines. Leaf tips are red. Before bloom the core of the plant turns fiery red-purple with tiny blue flowers.
- N. *cruenta* forms a 24-inch upright rosette with straw-colored leaves edged with red spines. A most unusual neoregelia that needs full sun.
- N. *johannia* is a durable 20-inch species with a lavender center. Although not as handsome as others in the genus, it is still worthwhile.
- N. 'Marmorata' is a hybrid with yellow-green and crimson leaves; rosette to 30 inches across. Spring or winter flowering, it needs good light for proper leaf color. Flowers are white and deep in the cup. This one thrives on neglect. It's for people who "can't grow anything."
- N. 'Painted Lady' forms a 24-inch rosette of dark-green leaves suffused with brownish-red markings; violet flowers.
- N. 'Purple Passion' grows to 24 inches across; fine purple-red leaves; small pink flowers deep in cup.
- N. 'Red Knight' makes a 20-inch rosette of handsome bright-green leaves heavily banded with maroon; flowers almost violet.
- N. *spectabilis*, the "painted fingernail" plant, grows to 30 inches

The following list gives you an idea of the variety of tillandsias available. No specific sizes are given, because plants vary greatly in size and many have recurving leaves, making it difficult to estimate exact proportions.

- *T. anceps* is a small, stemless species with numerous arching leaves and a large ovoid inflorescence, pale green or rose with blue petals.
- *T. brachycollis* is handsome, with many leaves. At blooming time the center foliage turns from green to coppery-red with purple flowers. Mount this species on a slab of tree fern.
- *T. bulbosa* has a bulbous base with narrow, leathery leaves. The inflorescence is magenta and white. An oddity best grown on a piece of bark or branch.
- *T. butzii* is small, with thin, twisted, cylindrical leaves that are purple-spotted. Rose-colored bracts have purple petals and yellow stamens. A pretty species that blooms in spring.
- *T. capitata* 'Giant Orange' makes a rosette of leathery grayish green leaves covered in silver with purple-red margins. At bloom time, foliage turns reddish and the inflorescence is a green crown.
- *T. caput-medusue* resembles *T. bulbosa*, small with vivid blue flowers.
- *T. caulescens* has gray-green spiny leaves and a handsome pendant inflorescence with red bracts and yellow flowers.
- *T. circinnata* is a small rosette with silvery-gray, leathery leaves and small lavender flowers on a flattened spike.
- *T. concolor* has stiff gray leaves in a star-shaped rosette. Bracts rose or green on erect spike. Tubular flowers are purple.
- *T. cyanea*, one of the most popular tillandias and rightly so, is a regal plant with graceful, arching leaves resembling a palm. From the center of this medium-size species, an erect flower stalk bears a feathery pink sword of large, purple-shaded flowers—a stunning bromeliad that needs moisture and humidity to bloom. This is one of the more difficult ones to grow but well worth trying.
- *T. dasylirilfolia*, a 20-inch blue-green rosette, has a branching flower stalk that is deep rose, petals whitish-green.

- *T. fasciculata* is medium size with blue or purple flowers. Among the many varieties available, leaf and flower color vary somewhat. A good one for the beginner.
- *T. flexuosa*, sometimes called *T. aloffolia*, has coppery-green twisted leaves with silver crossbands. It bears red bracts and white flowers.
- *T. geminiflora*, a dense rosette of green leaves, has a branched inflorescence and is a good plant for a bromeliad tree. An easy species and most decorative.
- *T. ionantha* is a dwarf, hardly more than 2 inches high, but it will astound you at blooming time in spring when all the leaves blush fiery red and tiny, pretty purple flowers appear. Grow it in sun or bright light; it needs little care.
- *T. juncea* is small and pretty, with narrow leaves in a tufted growth and a red flower crown. A favorite of mine.
- *T. leiboldiana* has arching, straplike, wide, gray-green leaves, spotted with maroon; red bracts and violet flowers at bloom time.
- *T. lindenii* is similar to *T. cyanea*, with long, graceful, tapered, reddish-green leaves. The inflorescence is blue. A real beauty but difficult to bloom.
- *T. paraenis* is a good windowsill plant. Leaves are grayish-green, the flower bracts pink.
- *T. punctulata*, with narrow, silver-gray, pointed leaves, bears a heavy inflorescence densely set with rose-red bracts and purple-and-white flowers.
- *T. streptophylla* is medium sized with a bulbous base and grayish-green, curving foliage. The inflorescence is branched, almost the same color as the leaves, with pink bracts.
- *T. stricta* is a rosette of recurving, leathery leaves covered with silver scales; rose bracts and blue flowers.
- *T. tricolor*, with grayish-green leaves edged red, is of medium size. The pink-and-red inflorescence is upright and branched, rising well above the plant.
- *T. xerographica* is a stiff rosette with narrow, concave, recurring silver-gray leaves; branched inflorescence with rose bract leaves; flattened lateral greenish spikes on red stems. The petals are purple.

- *V. platynema variegata* is a 40-inch rosette of blue-green leaves; the inflorescence is featherlike with purple bracts and greenish-white flowers.
- *V. regina* forms a 48-inch rosette of handsome green, waxy leaves speckled at the base; the tall inflorescence is branched with rosy bracts.
- V. 'Rubin', a recent hybrid, has an open rosette to 14 inches, green, glossy leaves, fiery red bracts, and a flower crown with yellow petals.
- *V. splendens* grows about 12 inches and is a perfect houseplant. The green foliage is mahogany striped and the thin, thrusting spring and summer inflorescence is orange colored.
- *V. s.* 'Meyer's Favorite' grows to 48 inches across, a glowing rosette of apple-green leaves blotched dark green; tall, fiery crown.
- *V. s. mortefontanensis* grows to 24 inches across, with handsome green foliage banded with brown; tall, typical "flaming sword" flower spike.

Sixteen

ORCHIDS

Contrary to popular thinking, not all orchids thrive in warm temperatures and high humidity. Orchids can be grown in three different climates, depending on the species: warm, temperate, and cool. In Florida, orchids that prefer warmth (75 to 90 degrees F by day, 10 degrees cooler at night) and those that prefer temperate climates (60 degrees F by day, 10 degrees cooler at night) will prosper; here I discuss the best ones and tell you how to grow them to your satisfaction.

Aerides

Aerides are epiphytic orchids native to tropical Asia. This genus of 60 species varies in size: some are several feet tall, others are dwarf (to 12 inches). The waxy-textured flowers usually are scented. Give aerides sunshine but not direct midday sun. Plants do not like their roots disturbed, so grow them 3 to 4 years before repotting them. During this time, merely resurface the soil with fresh compost when plants put forth new growth. Give these orchids moisture, daytime temperatures of 62 to 78 degrees F and 60 percent humidity during the summer. While plants are actively growing, spray them with a fine mist once a day. During the winter, decrease watering, but do not let plants become so dry that their leaves shrivel.

many are sweetly scented, and the inflorescence remains 2 to 4 weeks. In the summer, grow the plants in a western exposure, so they can get 3 to 5 hours of afternoon sun. In the winter, move plants to a southern or eastern exposure. Repot yearly in fir bark after flowering or before new growth appears; pot tight.

Let cattleyas dry out thoroughly between watering year-round. After flowers fade, let plants rest 5 to 7 weeks without water (but occasionally mist them). During the resting period, 58 to 62 degrees F at night is best. Cattleyas do well when summered outside.

C. *aclandiae* bears one or two 4-inch diameter olive-green flowers blotched brown-purple; the lip is magenta with darker purple. Grow on a slab of tree bark rather than in a container. C. *citrina* has bright yellow cup-shaped flowers 2 inches across. I grow this pendant plant on a block of wood. C. *dolosa* has magenta flowers with a yellow disk in the lip; it is winter-flowering. C. *forbesii* has two to five greenish-yellow flowers 3 1/2 inches across; the yellow lip is streaked red on the inside. C. *nobilior* has rose-colored flowers. C. *luteola* is only 6 inches tall, with 2-inch pale yellow flowers with a white lip; the sides are streaked purple. C. *o'brieniana* has one to three large rose-colored flowers. C. *schilleriana* has dark rose-brown flowers 4 inches across; the dark rose lip is edged with pink. It blooms in the late summer. After pseudobulbs mature, let plants rest bone-dry 5 to 7 weeks. C. *skinneri* is about 2 1/2 feet tall and bears two to eight rose-purple flowers about 3 inches in diameter.

Dendrobiums

This is one of the largest genera, with more than 1,500 species. Dendrobiums fall into five groups: (1) pronounced pseudobulbs, (2) evergreen cane-type pseudobulbs, (3) deciduous cane-type pseudobulbs, (4) evergreen cane-type *Phalaenopsis* hybrids, and (5) black-haired short-stemmed plants. The flowers in groups 1 to 4 last several weeks.

Group 1 (pronounced pseudobulbs) is easy to grow and is recommended for beginners. Plants need about 4 hours of sun daily and abundant watering until growth is mature. To encourage flower spikes, give plants a 3- to 4-week rest without water after growth matures. After plants flower, let them rest 5 to 7 weeks without water. Repot every other year in 4- or 5-inch containers. Temperate to tropical conditions suit most dendrobiums.

D. aggregatum is a dwarf, to about 10 inches, and bears small, scented, vivid yellow flowers in the spring. *D. densiflorum* grows to 20 inches in height. The drooping spikes bear many 2-inch golden yellow flowers in the spring. It often blooms from old as well as new bulbs.

Group 2 dendrobiums, with evergreen cane-type pseudobulbs, usually are big, sometimes taller than 6 feet. The flowers are perfectly arranged, like a bunch of grapes, and exceedingly pretty. These orchids need dappled sunlight and even moisture year-round, except just after flowering, when water should be reduced a bit for about 1 month. Flowers appear in spring or early summer.

Repot every other year in fir bark in 4- or 5-inch pots. Maintain daytime temperatures of 72 to 80 degrees F, 58 to 64 degrees F at night. If you are unable to bring these *dendrobiums* to bloom, rest them for about 3 weeks after growth has matured and move them closer to the window, where the 5- to 6-degree drop in temperature may induce bud formation in the winter.

D. dalhousieanum has 5-inch-diameter tawny yellow flowers; they are faintly shaded crimson. *D. densiflorum* has deep yellow flowers. *D. thyrsiflorum* bears bunches of crystal-white flowers that have an orange lip. This is a magnificent species.

Group 3 dendrobiums have deciduous cane-type pseudobulbs. They are also called *nobiles*. They produce two or three large and delicate pink, lavender, and orange flowers from the nodes along the top of each bare cane. While in growth during the summer, the plants need moisture and warmth; when foliage has fully expanded (a solitary leaf rather than a pair of leaves), stop watering—this is usually in October. Move plants to an unheated part of the home where the nighttime temperature is 48 to 55 degrees F. Do not water plants at all during the winter.

When buds start appearing as swellings along the nodes, move plants back to where the nighttime temperature is 58 to 64 degrees F, and resume watering as buds become larger. Repot these orchids every other year in 4- or 5-inch containers.

D. fimbriatum sheds its foliage every other year. This pretty little *dendrobium* has brilliant orange flowers. *D. nobile* has white flowers tipped purple, with a dark crimson blotch in the throat. This species has many good hybrids of various colors. *D. pierardii* bears paper-thin, 2-inch-diameter, blush-white or pink flowers veined rose-purple. It is quite dependable. *D. superbum* bears lots of large lilac-colored flowers

atropurpureum has brown-and-pink flowers with a red-striped lip in the early spring. It likes the same temperatures as *E. aromaticum*. *E. vitellinum*, a dwarf, grows to about 8 inches; it has brilliant red flowers. It likes 54 to 60 degrees F at night.

Group 2, stemlike pseudobulbs, has a habit similar to that of the cattleyas. These large plants grow slowly. Plants need filtered sunshine and do best when potted slightly loose in fir bark mixed with sphagnum. They need even moisture year-round, with a few weeks of rest after flowering. These epidendrums are slow to come back with new growth; do not force them. These species like about 70 degrees F.

E. prismatocarpum bears bright yellow blooms blotched vivid purple. This showy plant really catches the eye. *E. stamfordianum* has erect scapes of brilliant yellow flowers spotted red. The delightfully scented inflorescence lasts a fairly long time. The plant does best with a complete rest of 5 to 7 weeks (no water) after flowering.

Group 3 is made up of the reed-stem epidendrums, those without pseudobulbs. They are known for their constant blossoming. The plants need tons of water year-round and a few hours of west or south sun. They must be repotted every year.

E. o'brienianum has 1-inch flowers clustered at the top of the plant. As the lowest flowers fade, new ones appear at the top. Colors range from pink to lavender to brick red. This plant does well in the ground outdoors in a mild climate (grow in terrestrial compost).

Oncidiums

This genus has more than 700 epiphytic species. The spray orchids produce long spikes of yellow flowers marked brown. The plants have compressed pseudobulbs; some are almost without pseudobulbs; and others have pencil-like leaves. All oncidiums are evergreen. Usually the flower spike is flexible and arching, up to 5 feet in length. The small flowers appear in large numbers or sparsely, depending on the species. Many oncidiums bloom in the autumn or winter, and the flowers last 7 to 9 weeks.

Most of these plants need full sun for good flower production. Repot oncidiums every other year in well-drained fir bark mixed with chopped tree fern. While actively growing, the plants need plenty of water and generally high humidity (50 to 70 percent). After new growth has finished, plants must rest bone dry for 2 to 5 weeks. After flowers fade, plants again need another rest for several weeks. Oncidiums must be summered outside.

O. ampliatum has turtle-shaped pseudobulbs and bears small, spray-type yellow and red-brown flowers. When flowers fade, cut the same spike below the last node: The plant may produce a second node. *O. leucochilum* has yellow-green flowers barred brown. *O. ornithorynchum* is about 14 inches tall, with hundreds of tiny lilac-colored flowers. It needs semishade and a nighttime temperature of 52 to 58 degrees F in the winter. *O. sarcodes* grows to 16 inches. The pretty scalloped flowers are yellow and chestnut-brown. *O. splendidum* has solitary, cactus-like, 12-inch-long leaves. The flowers are a vibrant yellow barred brown; the large and broad lip is yellow. *O. wentworthianum* has yellow flowers blotched brown.

Phaius

Phaius spp. are deciduous terrestrial orchids from China, Africa, and Madagascar. They lose their leaves during the second year. Most phaius adapt to varying temperatures. Leaves often are longer than 4 feet; flower spikes are erect. A healthy phaius can bear 10 to 20 large flowers; flowers are long-lasting, scented, and sometimes as large as 5 inches in diameter.

Phaius need filtered sunlight, not direct sun, so a western exposure is ideal. Repot plants every other year in a compost of loam, manure, and fir bark, in 8- to 10-inch containers. Apply heavy watering and feeding until the flowers actually open, at which time give plants a good rest in a cooler temperature. While plants are resting, dry out the compost slightly, and do not mist the foliage at all. Watch out for thrips, which love these orchids.

P. grandifolium (*P. tankervilliae*) bears flowers that first are pale and then later darken considerably. The sepals and petals are yellow-brown and silver-hued; the lip is rose-purple and whitish with a blue spot in the center. *P. maculatus* has buff yellow flowers; the lip is marked red on the front lobe. The pretty foliage is spotted yellow.

Phalaenopses

The *Phalaenopses*, or dogwood orchid, include about 70 species. The white flowers of the hybrids are popular for corsages and cutting. Plants

Seventeen

CACTI AND SUCCULENTS

Only the orchid family can compete with cacti and succulents when it comes to giving the exotics gardener a vast selection and rare beauty. Succulents are usually grown for their stunning foliage, attractive growth habits, or striking flowers. Cacti are prized for their unusual shapes and colorful flowers. The golden barrel cactus has beautiful yellow spines; the Christmas cactus bears exquisite red or pink flowers; and large cacti, such as the cereus, add structural drama to areas. Euphorbias are bizarrely contorted; echeverias have rosettes of leaves that look as if they were carved from stone. Agaves have leaves that look like giant tufted flowers.

Most cacti have shallow roots, and planting can literally be a pain (if a thorn gets you!). If the cactus is large, protect your hands by wrapping the plant in newspapers before planting, and always wear gloves. Dig the hole to match your plant; if it is large, dig down 16 to 20 inches and provide plenty of horizontal space—make sure the diameter of the hole is at least as wide as the width of the plant. Secure the plant by firming the crown; the cactus should neither tilt nor lean. If you need some help keeping it upright, use small hunks of wood placed at the base in a triangular arrangement to prop up the cactus until its roots take hold. Large specimens can be difficult to plant, but smaller cacti and succulents should not present a problem. Both cacti and succulents generally need good watering during the summer, less during the winter.

Cacti

Aporocactus is a species of unusual chain-type cacti not often seen but worth growing. Plants bear dozens of small red flowers in the winter. It can tolerate cold to 40 degrees F. *A. flagelliformis* is the best for Florida gardens; it has pendant leaves with reddish hairs.

Cephalocereus are noted for their tall and columnar or branching growth and a coat of long and woolly hair. Most of these cacti flower at night. *C. palmeri* (woolly torch cactus) is hardy and has short blue-green spines and tufts of white hair. *C. polyanthus* (Aztec column) is slow growing and has fluted ribs and yellow-brown spines. *C. senilis* (old man cactus) has ribbed and columnar growth and spines hidden among its long white hairs.

Chamaecereus are small cacti, with bright red flowers. Short shoots branch from the base, producing a clump effect. *C. sylvestri* (peanut cactus) has dense clusters of short green branches.

Cleistocactus are tall, slender, columnar, and generally large, with most flowers orange-red. The spines are white or red. *C. baumanii* (scarlet bugler cactus) has stiff stems topped with white spines and bright red tubular-shaped flowers. *C. straussii* (silver torch) has silver hair and dark red flowers.

Epiphyllum cacti are spineless, with large, saucerlike flowers in many colors. They are easy to grow in filtered light and a well-drained sandy soil. *E.* 'Conway's Giant' has scalloped green leaves and red flowers. *E. crenatum* has beautiful white flowers. *E.* 'Padre' displays pinkish-white flowers.

Opuntias are large or small, handsome or ugly. The flowers usually are yellow or pink. Plant only mature specimens. *O. basilaris* (beaver tail cactus) is upright and blue-green. It can withstand low temperatures. The flowers are pink to carmine. *O. bigelovii* (teddy bear cactus) is branching, with elongated stems covered with white hairs. *O. microdasys* (bunny ears) is a dwarf Mexican species. The spineless pads are covered with tufts of golden bristles. *O. vestita* (old man opuntia) has very hairy and small deciduous leaves. It bears red flowers.

Rhipsalis are trailing plants with small but showy flowers. *R. paradoxa* (link cactus) has flat green leaves with sawtooth edges. The tiny flowers are white.

Two genera, *Schlumbergera* and *Zygocactus*, are known as Christmas or Easter cacti. *Schlumbergera* have white, cream, orange, red,

or fuchsia flowers. Give these cacti lots of sun in the fall and winter. S. *bridgesii* has pink or red terminal flowers.

Zygocactus has stunning flowers of pink or red. It needs even moisture year-round, good air circulation, and 6 weeks of darkness in the late fall to force flower buds. Z. *bridgesii* (the true Christmas cactus) has bright red or rose-red flowers and trailing, flattened stems. Z. *truncatus* is the Thanksgiving cactus.

Succulents

Agaves and aloes are the most common succulents used as landscape subjects. But overall, there are endless plants to select for your outdoor scene.

Agaves are slow-growers that can be small, to 10 inches, or giant, with 5-foot-long leaves. They are easy to grow. A. *americana marginata* (century plant) is trunkless and grows to 5 feet. The green leaves are edged yellow. A. *attenuata* grows to 3 feet and has soft gray-green rosettes. A. *filifera* (thread plant) has narrow olive-green leaves that bear loose, curled threads at their margins. A. *horrida* grows to 20 inches in diameter. The dark green leaves have sharp marginal spines. A. *medio-picta* reaches 5 feet. Green leaves have yellow center stripes. A. *miradorensis* (dwarf century plant) grows to 3 feet. A. *picta* has pale green leaves with white margins and small black teeth. A. *victoriae-reginae* is beautiful, with olive-green leaves penciled with white edging.

Aloes—mostly African natives—have a rosette growth habit and fleshy, sword-shaped leaves. These succulents are members of the Lily family. The flowers usually are orange. Aloes prefer shade and not too much water. Once established, they will almost take care of themselves. A. *arborescens* (candelabra aloe) can reach 10 feet. The thick leaves are blue-green. A. *cilaris* has soft, toothed gray-white leaves 6 inches long. A. *globosa* (crocodile aloe) has gray-green leaves. A. *nobilis* (gold-spined aloe) is a rosette, to 20 inches in diameter, with bright green leaves edged yellow. A. *polyphylla* (spiral aloe) grows to 20 inches in diameter, with gray-green, sickle-shaped leaves, which are dark brown at the tips. A. *striata* (coral aloe) has a short stalk and pointed, gray-green leaves with a narrow pinkish edge. A. *variegata* (partridge breast; tiger aloe) has green leaves marked with white wavy bands.

Crassula have fleshy leaves and stems and tolerate a lack of moisture. Some have gray or blue foliage: others are green. Several have branching stems while still others produce low rosettes. Plants like bright light and ample water. Used mainly as background plants in warm climates where they grow quickly. C. *argentea* is the most popular species growing to 5 feet with fiery red flowers. C. *falcata* is also popular.

Echeverias are beautiful rosette plants appearing like sculpture in the landscape plan. Rosettes are colorful, some almost chalky white, others with bluish or pinkish casts. Orange or red tubular flowers appear in spring and plants are best used as background in gardens. E. *derenergi* and E. *elegans* are the most popular species but there are numerous hybrids available.

Euphorbias include the beloved poinsettia. All *euphorbias* are easy to grow and need sun and even moisture. E. *clavarioides* is columnar-shaped, with cupped yellow flowers. E. *lactea* has many branches. Each branch has three to four angles, spines, and a whitish central stripe. E. *milii splendens* (crown of thorns) has spiny stems, tiny leaves, and bright red flowers. E. *obesa* (basketball plant) is a stunning multicolored ribbed sphere.

Kalanchoes are easy to grow succulents. K. *blossfeldiana* (Christmas plant) is known for its red or orange-yellow flowers. K. *daigremontiana* has stalked, shiny green leaves that are flecked on the underside. K. *fedtschenkoi* has blue-green leaves clustered at the tops of stems. The flowers are brownish-pink. K. 'Rose Leaf' has velvety gray to green-brown leaves. K. *tomentosa* (panda plant) has white leaves covered with brown dots.

Sedums are low-growing succulents with thick or needlelike leaves. Some varieties trail; others are bushy. Give these plants sun and even moisture. S. *adolphi* (golden sedum) is a sprawling plant with short, fleshy, yellow-green leaves tinged with red. S. *multiceps* is a miniature tree. The flowers are yellow. S. *oxypetalum* is deciduous, with peeling, papery bark. It grows slowly. S. *sieboldii* cascades and has solitary stems covered with wedge-shaped leaves lined red.

Yuccas are a large group of wonderful plants often overlooked in Florida landscapes. They generally produce rosettes of green leaves and handsome flowers borne on tall stalks. Y. *aloifolia*, the Spanish bayonet, is very attractive as an accent plant, as are Y. *elephantipes* and Y. *filamentosa*.

Eighteen

Starting Your Own Exotic Plants

Plants are generous with offspring. In fact, sometimes I hardly know what to do with all my divisions, cuttings, and seedlings. Seeing your own plants grow brings satisfaction, and, as with your children, you like your own best. You do not have to be a botanist to propagate plants. It is an easy procedure, and there are many ways of going about it. And having your own supply of exotics is important if you cannot find plants in local nurseries.

Stem and Leaf Cuttings

The best time to make stem cuttings is usually spring or summer, when plants are growing actively. Select sturdy shoots. Cut off a 4- to 6-inch piece of stem with several pairs of leaves just above another pair on the mother plant. Trim off the cutting squarely just below the lowest pair of leaves and remove these leaves. Remove any flowers on the cutting as well. You can root cuttings of some plants philodendrons and syngoniums for example) in a jar of water on a windowsill. Others—the majority of cuttings—should be planted in vermiculite, sand, or other light growing medium, moistened, and placed in a warm, humid atmosphere. A breadbox, baking dish, plastic container, casserole—any container with a transparent cover—makes a good propagating case. You can also use bulb pans with glass jars over them or sealed plastic bags propped up so as not to crush the cuttings. Before planting, you can dip the base of the cuttings in hormone powder to stimulate root growth. Then insert them, 1/3 to 1/2 their length, in the propagating mixture. Grow cuttings warm, 70 to 80 degrees F, and place them in light rather than in direct sun. Keep the growing medium moist. Lift the cover of the propagating container for an hour or so a day to allow some ventilation.

Many plants—rex begonias, sedums, kalanchoes, and others—can also be propagated from leaf cuttings. Select a firm, healthy, mature leaf. Make a few small cuts across the main veins on the underside of the leaf with a sterilized knife or razor blade (run the instrument through a match flame). Lay the leaf on moist vermiculite in a propagating container. To ensure contact between leaf and growing medium, weight the leaf with pebbles. Plantlets form at the slit veins and draw nourishment from the mother leaf. When the plants are large enough to handle easily, put each in a 2- or 3-inch pot. You can also simply plant a single leaf as though it were a stem cutting.

Certain large foliage plants, such as alocasias, dieffenbachias, and philodendrons, can be propagated by a process called *cane cutting*. Sever a 4- to 8-inch piece of mature stem (or cane). Coat both ends with sulfur, lay it in a rooting medium such as sand, cover the cane lightly, and press it firmly in place. New plants will form from dormant eyes all along the cane.

Stolons and Offsets

Many plants produce stolons or runners. If you want to propagate a plant of this type—orchid or bromeliad for example—cut the 3-inch-long offsets of these plant-producing stems and handle them like other cuttings. Roots will develop and plants will be ready for potting in soil in four weeks to four months, depending on the type. Bromeliads, agaves, gesneriads, and many orchids develop offsets or suckers at the base of the mature plant. When these offsets are 2- to 4-inches long, cut them off with a sterile knife and root them as you do cuttings.

Root, Bulb, and Rootstock

Plants that make root clumps or multiple crowns—clivias, some orchids, and some ferns—can be propagated by division. Pull such plants apart into two or three sections, making sure that each section has some roots. If plants are massive, as philodendrons can be, or woody, as are some other aroids, divide them with a clean, sharp knife, preferably sterilized. Cut back the foliage of your divided plants to encourage new growth, and plant the divisions in pots 2- to 4-inches larger than the rootball or crown.

Bulbous plants, such as amaryllis, eucharis, and haemanthus, increase themselves by producing smaller bulbs: These can be separated from the mother bulb and potted up to make new plants. Sometimes when you are not watching, one of the smaller bulbs that develops within the mother bulb starts to grow on its own; this is called an offset bulb.

With some gingers and other tuberous plants, clumps of fleshy roots develop—all joined to one stem. To propagate these *rootstocks*, divide the stem into sections—each with an "eye" and attached to a root.

Air Layering

Plants with large leaves and woody stems (dieffenbachias, ficuses, and philodendrons) are often difficult to reproduce from cuttings and can instead be propagated by *air layering*. On a healthy, sturdy stem near the top of the plant either remove a strip of bark about 1-inch wide directly below a leaf node or cut a notch about half way through the stem. Then wrap a big clump of moist osmunda or sphagnum moss around the strip or notch and cover it with a piece of plastic wrap secured at top and bottom with string (the ball must be moisture-proof or growth will not start).

Air layering takes time. It may be six to nine months before roots form. When you can see them poking through the moss ball, sever the new plant just below the ball of roots. Then pot it, moss and all.

Fruit and Nut Trees

Living in Florida has advantages other than a great climate and a chance to grow a wealth of ornamental garden exotics. You also can cultivate tropical fruit trees from seeds or pits if time allows. It takes several years to maturity. These double-duty plants add handsome decoration to the landscape and provide you with delicious exotic fruits that are an eating adventure. If you are a cook be sure to try these wonderful tropical fruits to add zest and variation to ordinary recipes. Many exotic fruits are appearing in food stores, but they are expensive. With a modicum of care and proper selection, however, you can have your own treasure of taste with little cost.

You can start your exotic fruit trees from seed or pits (that's how I did it in northern climates under glass, but you can also buy young trees (finally) at some nurseries. The wonderful thing about exotic fruit trees is that even if they didn't give fruit, they would still make handsome additions to the southern landscape. Trees like the cashew nut, loquats, mango, and avocado all have beautiful foliage and form. (I have omitted citrus fruits simply because they cannot be classed as "exotics" in Florida.)

Avocado

The lush green meat of the avocado (*Persea americana*) has supplied many a table with salads, and its rather heavy rounded pit has furnished many households with a lush green plant. Part of the avocado's popularity is due to the fact that the pit is easy to get started. There are several methods of approaching the birth of your avocado. Some people (I'm one) merely clean the pit and put it in a glass jar half full of water. You can also prop the pit on toothpicks. In a few weeks it's off and growing roots, and can be planted in soil.

To be more certain of germination, however, cut a thin section from the apex and the base and peel away the papery pit coating. Put the pit in the soil or water base downward—that is, the broadest part of the pit. If in soil, don't embed the pit too deeply; cover it with about one-half inch of soil.

Be sure the avocado plant has sufficient drainage; although it isn't choosy about soil it is particular about stagnant water at the roots. The plant has a tendency to shoot straight upward, so once it is growing well, clip off the top to encourage side branching or it will get leggy and unattractive. Even so, you may have to stake the plant to keep it upright.

Repot the avocado frequently (every six months), each time to a larger pot. Eventually you can transplant it into the garden and you'll have a handsome tree.

Chinese Gooseberry

Don't let the name throw you; this is the kiwi fruit. The true Chinese gooseberry (*Actinidia sinensis*) is native to China, but the fruit we find at local markets is from New Zealand. This lovely twiner, with its fuzzy leaves, is ideal for trellis growing. The fruit itself is shaped like a grape, the size of a small egg, and is chartreuse in color. When sliced and sprinkled with lemon the fruit is somewhat puckery but good. Pick out the small seeds from the center of the fruit and dry them on a blotter or newspaper.

Because the kiwi is subject to damping off, sow the seed in a coffee can or flat azalea pot; cover the surface of the soil thinly with vermiculite to protect the susceptible stem-to-ground junction. It takes about eight weeks for kiwi seed to germinate. If you have trouble sprouting the kiwi pits the first time around, give them a cooling-off period in the refrigerator before the second attempt. This is known as stratification. To do this, mix the dried seed with some sphagnum moss in a plastic bag and store at 45 degrees F for about forty days. (This simulates the natural germination cycle of the kiwi seed in nature.) Then replant the seed in containers. You should have sprouts within three weeks. Transplant them when they're a few inches high.

In the home landscape, plant the kiwi where it will receive full sunlight and give it copious amounts of water. The kiwi prefers temperatures of 78 degrees F in the day and no less than 45 degrees F at night.

Citron

I found this delightful fruit in Chinatown during the winter season. Citron (*Citrus medica*) is, under ideal growing conditions, a dwarf tree of about eight feet with a large (6- to 8-inch-long) ovoid fruit and bright green toothed leaves. The fruit, which is rough-textured and fragrant, provides the citron peel used in fruit cakes. Meticulously peel the citron and reserve the peel for future use. Wash away all pulp and plant the seed or pit in sandy soil in a shallow container. Bury the seed about one inch deep in the soil and keep warm (72 degrees F). The sprouts should start in about a month. When about 4 inches tall, put the seedlings into individual pots with soil that contains some calcium. The citron is an evergreen tree, so keep watering it through winter.

Date Palm

Dates (*Phoenix dactylifera*) are infinitely good for you and loaded with all kinds of vitamins, and date palms are lovely trees, so there are two reasons not to dismiss this venture with a shrug of the shoulders. If you want to start your own date palm, don't make the mistake of using pasteurized dates that have been steamed and preserved with chemicals.

Get raw dates, that haven't been tampered with, they are usually available at health food stores. Break open the date and wash several pits. Plant them 1 to 3 inches deep in a starter mix. The time for germination varies, but it could be as long as two months, so don't give up in disgust. Keep the container in a warm place with good humidity. (Placing the entire pot inside a closed plastic bag is a good way to maintain high humidity.)

In most plants the seed cotyledon that acts as a reservoir for food usually emerges from the top of the seed and sprouts directly out of the ground—not so with the date. It comes from the bottom and travels through the soil (like a root), coming up many inches later. When this root is about an inch long above the soil, it is time to transplant to a large tub of rich soil so the plant can grow on. In a few weeks the sprout will be joined by another, and presto, fronds! Your date is on its way. Give the plant plenty of sun, good moisture, and occasional feeding to keep it growing.

Fig

The fig (*Ficus carica*) is a deciduous shrub or small tree belonging to the mulberry family. Its lovely lobed green leaves make it valuable in the landscape even without its fruit. The fig makes a splendid patio or indoor plant that grows with little attention, save for constantly moist soil and bright light.

Fig tree seedlings are available through most mail order nurseries, and well-grown saplings can be found in some local nurseries. To start your own fig tree, take a cutting from a mature tree and start it in vermiculite. In about 6 to 8 weeks, the cutting will take root, enabling you to transplant it into the yard.

Guava

You can buy a guava (*Psidium guajava*) plant at nurseries, or grow it from the pit. Called the sand plum by the Aztecs, the guava is delicious eaten fresh or as a dessert with cream. The hard pits must be clean of the fleshy fruit, so after eating wash the pit in warm water and get it started right away. Cover the pit with about 1/4 inch of soil and keep it in a warm, bright place. The first shoots should appear in two to four weeks; if not, try again with fresh pits. The tree grows in an upright manner and can be pruned and trimmed without harm. The guava is a decorative outdoor garden or patio plant where temperatures don't fall below 55 degrees F. Once it is actively growing, remove the suckers that form at the base. Prune out all but three or four good branches.

Litchi Nut

You've eaten these no doubt in Chinese restaurants, where they're served either within a dish (succulent and sweet) or dried (sweet and dry). From southern China, the litchi (*Litchi chinensis*) has been cultivated for more than 2,000 years. You can purchase small trees at specialty nurseries or start your own from seed. To do this, you'll need fresh litchis; the dried ones won't grow. The smooth-skinned pit, dark brown-black in color, is pretty in itself without planting, but once planted it produces a lovely narrow-leaved plant. Litchi is difficult to get started, so sow several seeds, one to a pot, in an acid-rich soil. Plant them about 1/2 inch deep. Keep them shaded and water them well. If you don't see signs of growth in two to three weeks, try again with fresh seed. Keep litchis well watered and gradually expose them to bright light, but never direct sun.

Loquat

The loquat (*Eriobotrya japonica*), another Asian plant, should really be used more; the fruit is especially good just eaten out of hand or in poultry casseroles. The plant, too, deserves more attention because it's a lovely bold-leaved green beauty. The orange fruit resembles an apricot when it is ready for picking because of its orange color. Wash and dry the seeds and plant them 1/2-inch deep in good fertile soil that has ample drainage. With proper care and frequent repottings, the loquat can grow into an attractive tree in gardens where temperatures don't go below 25 degrees F. Loquats are sometimes called Japanese medlars.

Mango

The lovely mango was cultivated in India 4,000 years ago (that should impress guests), and the plant itself is impressive, with leafy green foliage. The fruit, delicious for breakfast or dessert, is yellow and red with black specks, and generally kidney-shaped.

There are several different varieties of mangoes (*Mangifera indica*), and several ways of starting this plant. A pit can be started either by drying it first for a few days, by soaking it in water for a few days, or by nicking its edge with a knife. In any case, first clean the pit by rubbing it with a stiff brush. (Wear gloves if you have sensitive skin—the mango does not like to be fondled; it can cause a rash.) Set the pit on end with the "eye" up and suspend it (as you would an avocado) with toothpicks in water or in starter mix so that the bottom inch or two is in the medium. Put the container in a warm, bright place; sprouts should appear in a month. Transfer the seedling to rich soil outdoors. Firm the soil around the collar of the plant but do not bury the stem. Keep soil evenly moist.

Papaya

The delicious fruit of the papaya (Carica papaya) is good for you as a dessert or as a digestive aid (it contains the enzyme papain, which is essential to digestion). Papaya flowers can be male, female, or hermaphrodite. In starting a plant, first wash the seeds well and remove the slippery outer coating; then sow immediately. Cover with a scant 1/8-inch of soil and keep in a warm, bright place. Germination should occur in four to eight weeks. You can also start seeds in vermiculite; keep a plastic bag over the pot to maintain high humidity. Papayas are prone to damping-off. Water copiously in the summer but less frequently in the winter. Repot the plants frequently, the first time six months after you start, and then again six months later. The papaya's leaves are scalloped and seven-lobed and remind me of the popular houseplant called the umbrella tree.

Passion Fruit

There are more than 400 species of *Passiflora*, but only some provide the fruit used for jellies or desserts. My rampant, lovely passion fruit vine was of a nonedible variety, but it still produced a lavish plant. All passion fruit vines produce leaves that are large and deeply scalloped. Passion fruit can be started from seed as soon as the fruit (edible or not) is available. Plant seed in a light soil and provide high humidity and warmth (78 degrees F). Germination should start in a month or less. When the seedling is 2 to 3 inches tall put it in an individual pot. Keep it in a bright place with moderate temperatures; avoid extremes.

Persimmon

If you've ever seen the lovely persimmon (*Diospyros kaki*) tree, you'll want one. It has ovate green leaves, and its bright orange fruit is eaten fresh or cooked. Germination is erratic and may take from 2 to 10 weeks, depending on how you handle the seed. For best results, put the seed in some sphagnum moss after you've eaten the fruit. Store the seed and moss in a closed bag in the refrigerator for ninety days. Then sow the seed, covering it with a shallow layer of soil. Give it warmth and bright light and keep your fingers crossed.

Pomegranate

Many people are probably familiar with the small pomegranate, *Punica granatum nana*, sold as a houseplant. But you can also grow the standard pomegranate, *P. granatum*, from seed. Introduced into England from Spain in the sixteenth century, this is a delicious fruit and makes a handsome plant.

The fleshy coating of the pomegranate seed is eaten; it is cooling, sweet to the taste. Once you have eaten the goodies, take the seed and allow it to dry a few days. (You can also separate seed from pulp by running the fruit through a sieve and rinsing away the fleshy part.) Germinate the seed in shallow pans of vermiculite covered with plastic to assure good humidity. Mold may accumulate on the seed, but don't panic—this is a symbiotic relationship necessary to germination. In about two months the seeds will crack open and start growth. Transfer the seedlings to individual pots of rich soil—equal parts potting soil and humus. Cover the seed and the taproot with the soil and in a few weeks leaves should appear. Give the plant warmth and sun; the soil should be evenly moist, never soggy. The pomegranate dislikes high humidity.

Although starting the seed of this plant is very possible even for the novice, getting the plant to bloom is another. So be content with the lovely foliage; it is a delightful plant.

Rambutan

The rambutan (*Nephelium lappaceum*), native to Malaysia, is beginning to appear in specialty markets. It belongs to the same family as the litchi and is grown in the same way. The fruit is the white fleshy aril surrounding the single seed. It is sweet and acid and can be eaten raw or stewed. The seed has to be started in high humidity and good warmth in a sandy soil mix kept evenly moist. When germination occurs (and this may take many weeks) and green growth shows, transplant the plant to a pot of rich soil. Keep the seedling evenly moist and warm. The rambutan is a leafy branching plant that makes a distinctive accent. Although you might have to search for the fruit, it is worth the time.

Nineteen

Exotics as Cut Flowers

Exotic plants used as cut flowers can add much beauty to interiors. Tall stalks of ginger, feathery sprays of oncidium orchid, or the elegant orange blooms of heliconias in a vase of water, all cut from your own garden, will rival the arrangements found in florist shops.

Cutting your outdoor flowers for indoor decoration will not harm plants. But it is wise to remember that a cut flower is a living thing and although separated from the mother plant, it still needs nourishment and care. If you want to keep flowers and greenery fresh indoors as long as possible some of the following hints will help you maintain their beauty.

Different plants require different handling techniques because plants have different types of stems: woody, hollow, reedy, stout, and so on.

When cutting flowers remember that once cut they need water as quickly as possible. You can carry a small jug of water with you when cutting, or just cut and carry the flowers into the kitchen and submerge the cut ends in warm water as soon as possible. If you cut flowers and leave them out of water for an hour, you are doing them an injustice.

It is best to avoid cutting flowers in the heat of the day (the old wives' tale about morning cutting is valid). Cut flowers in early morning or at dusk. In the early morning and evening the plant's stems are filled with water and in better condition to survive cutting.

As you go to cut flowers, carry with you a suitable basket or container to put flowers in; it is difficult to cut flowers with one hand and hold them in the other. Be sure you have sharp knives or scissors because you want to make lean cuts, never ripping gashes. If you crush the capillary vessels in the stems, they cannot take up water later. Sterilize cutters occasionally to help prevent spreading bacteria to the flower stem and plant. I sterilize my cutters by running a match flame over them. You may prefer to be more professional and dunk them in a sterile bleach solution, which is fine.

When you cut flowers select those that are in loose bud stage; the buds should show flower color.

Always cut stems at a slant to expose more of the stem surface to water. And always cut the stem below a node; cutting at midpoint can weaken the plant. Handle flowers gingerly: Do not smash them around or treat them like a sack of potatoes. True, they can take some mishandling, but why stretch your luck? Flowers are fragile when cut, so handle them carefully.

When you get flowers cut and in water, soak them to their necks at room temperature for about two hours. Then move the flowers and container to an airy, cool (65 degrees F) place, overnight if possible. In a cool spot at night, plants transpire little; thus, stems and leaves stay crisp and filled with water.

When you are ready to arrange your flowers, recut the stems. Plants with woody stems should be split (from the bottom, 6 to 8 inches), and you can slightly smash these stems with a tap of a hammer. Immerse hairy-stemmed plants in tepid water. Recut hollow-stemmed flowers under water. Plants with oozy stems should be seared with a match flame.

Recutting stems under water may sound silly, but it does help preserve the quality of some flowers. Air bubbles can form during the brief period it takes to cut the flowers. The bubbles form because the crushed stems cannot take moisture. If you slice off a 1/4-inch of the stem under water it prevents a new air bubble from forming.

Before flowers are arranged, strip all leaves below the water line, because foliage decomposes rapidly. Most flowers are not fussy about the quality of water, so regular tap water is fine. A few chips of charcoal will keep water sweet and odorless. There are packages of prepared chemicals that come with cut flowers from florist shops; use them if you like. Quite frankly, I have never seen any difference between flowers in plain water or flowers with chemicals added to the water—the flowers in either instance last the same time.

Hints for Cutting Specific Flowers

Through the years I have found that different flowers respond to different treatment; so here is a rundown on how to cut and handle some favorite exotic flowers.

Curcuma Cut flowers when they are nearly open, then place them in warm water. Recut stems under water. Flowers can last two weeks.

Alpinia Cut flowers when they are three-fourths open. Split stems in warm water; condition overnight. Lasts two weeks.

Plumeria Cut at any stage. Submerge in cold water for a few hours and then remove. Cut stems again under water. Lasts two weeks.

Hedychium Cut flowers when half open. Condition in cold water overnight. Lasts two weeks.

Costus Cut when flowers are almost open. Recut stems at a slant under water, and let stand in cold water for one hour or so before arranging in tepid water. Lasts for over two weeks.

Heliconia Cut when flowers are almost open. Place in warm water. Stems can be recut after about five days and revitalized in warm water. Lasts two weeks.

Zingiber Cut when half-flower spikes shows color. Condition overnight in cool water. Lasts about ten days.

Epidendrums Cut when in bud, then place in warm water in a vase overnight. Flowers last about one week.

Kampferia Cut flowers when they are fully open. Recut stems under cold water. Lasts five days.

Oncidiums Cut when three-quarters open, in late afternoon. Cut stems so that two leaf nodes remain on plant; cut just

	above this second node or eye. Submerge stem in cold water overnight if possible.
Lycoris	Cut when flowers are almost fully open. Condition in cold water overnight if possible. Lasts about ten days.
Haemanthus	Cut when nearly open. Soak stems in cold water. Lasts about a week.
Cattleya	Cut when flowers are almost completely open. Submerge in cold water for a few hours. Lasts about ten days.

Keeping Flowers Fresh

Many times, after people cut flowers and put them in a vase of water, they forget them. Do not do this. Cut flowers, like any living plant, absorb water, and water evaporates. Replenish water daily, and your flowers will last a long time. After several days you might want to recut stems to a sorter length and rearrange the flowers; this will prolong the flowers' lives by several days. Finally, you can use the flowers one last time by floating them in water.

To keep flowers fresh, keep them out of drafts; they last significantly longer in a quiet place. Flowers will also last longer in a cool rather than a heated room. Also, it is best to keep flowers away from direct sunlight, which can wilt them.

Flower Arranging

Arranging flowers is an art in itself, and it is difficult to cover it in a short space; here we give the rudimentary elements of flower arranging. There are many books on flower arranging at libraries to help you.

You can learn a great deal about flower arranging by observing pictures and the professional arrangements you see in flower shops. There are some basic rules of design: (1) Scale—the size of the flowers must relate to each other harmoniously: Large, medium, and small flowers should be placed strategically to create a whole; (2) Balance—the arrangement should lean neither left nor right; this is achieved by repeating a flower several times in the total arrangement; (3) Proportion—this refers to the vertical and horizontal aspects of a bouquet; tall flowers should be balanced with a low mass of flowers.

An arrangement of exotic flowers: *Heliconia stricta*, Zingiber, and orchids. (A. R. *Addkison*)

The bird of paradise's dramatic lines make it suitable for pared down arrangements—or just displayed on its own. (*Barbara J. Coxe*)

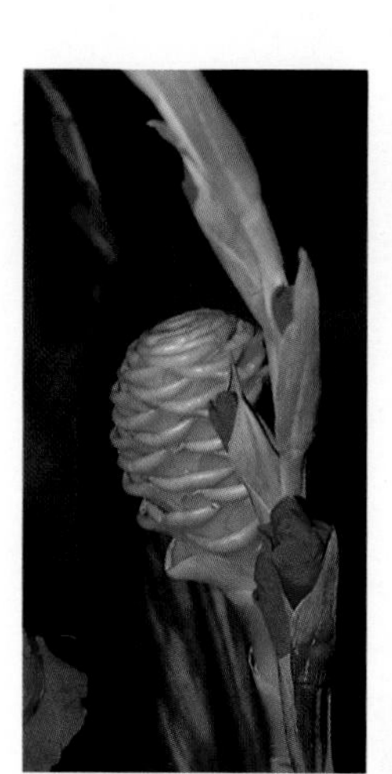

Zingiber spectabilis makes a fine cut flower and lasts for weeks with its unusual flower head. (A. R. *Addkison*)

Vallotas make fine cut flowers, and plants can be grown easily from bulbs. (A. R. *Addkison*)

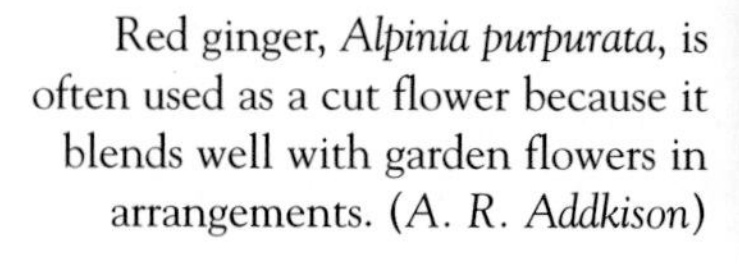

Red ginger, *Alpinia purpurata*, is often used as a cut flower because it blends well with garden flowers in arrangements. (A. R. *Addkison*)

In addition to the well-known vertical or horizontal arrangements, where the emphasis is either of height or mass (width), there are sophisticated designs that use a cascading effect and some that are fashioned on an arc. These artistic arrangements are simple, using few flowers, but require strategy to arrange; the technique is derived from Japanese methods. Of course, there still remain the lovely, informal, old-fashioned arrangements, where a great many flowers are used and a mass of color is created.

The selection of color for an arrangement is also very important. Colors must complement each other, and there should not be jarring effects such as blue flowers next to orange ones. There are no rigid rules on flower color in arranging; what you use depends on what you like. I have seen appealing arrangements made with ten different colors, but always the flowers were well-placed for balance and harmony throughout.

When you are arranging flowers, always consider the container as part of the total piece. Generally, a simple glass container or vase is best for most flowers, and its size should be in keeping with the size of the arrangement. Two dozen flowers crammed into a tiny vase will not be pretty; they will just look crowded. Six flowers in a huge container will appear uncomplicated but the overall effect is incongruous to the eye.

To select the proper size container, keep in mind that the container should be half as tall as the arrangement itself and half as wide as the span of the flowers. There are always exceptions of course (a single tulip in a glass), but generally the above guidelines are feasible for most bouquets.

To anchor the arrangement in a vase, many people us frogs (wire or plastic receptacles) to hold the flowers in place or florist's clay. Start by inserting flower stems in the center and work from left to right after you have a small grouping of blooms at the center.

Appendices

Gardens to Visit

Florida has excellent public gardens that feature exotics. Be sure to call or write before visiting these gardens; admission fees vary.

Bok Tower Gardens
1151 Tower Boulevard
Lake Wales, FL 33853
941-676-1408
157 acres of cultivated land. Offers classes. Hours are 8 A.M. to 6 P.M. daily. Admission closes at 5 P.M.

Butterfly World in Tradewinds Park
3600 West Sample Road
Coconut Creek, FL 33073
954-968-3880
Tropical gardens and a butterfly museum. Classes on butterfly gardening are held monthly.
Hours are 9 A.M. to 5 P.M. Monday through Saturday and 1 P.M. to 5 P.M. Sunday.

Fairchild Tropical Gardens
10901 Old Cutler Road
Miami, FL 33156
305-667-1651
At 83 acres, this is the largest tropical botanical garden in the continental United States. The world's finest collections of palms and cycads, along with flowering trees, shrubs, and vines.
Hours are 9:30 A.M. to 4:30 P.M. every day.

Flamingo Gardens and Arboretum
3750 Flamingo Road
Davie, FL 33330
954-473-2955
Sixty acres in what was once an orange grove. Heliconias are a specialty.
Hours are 9 A.M. to 5 P.M. daily.

Harry P. Leu Gardens
1920 North Forest Avenue
Orlando, FL 32803
407-246-2620
50 acres of gardens, including camellias, orchids, palms, cycads, and the largest rose garden in Florida. Offers classes year-round.
Open 9 A.M. to 5 P.M. daily.

The Kampong of the National Tropical Botanical Garden
4013 Douglas Road
Coconut Grove, FL 33133
305-442-7169
Former home of famous plant explorer David Fairchild.
Tours by appointment.

Marie Selby Botanical Gardens
811 South Palm Avenue
Sarasota, FL 34236
941-366-5731
Eleven acres open to the public, with a glorious tropical display house, strolling grounds, and a botany museum.
Hours are 10 A.M. to 5 P.M. daily.

Mounts Botanical Gardens
531 North Military Trail
West Palm Beach, FL 33415
561-233-1750
This botanical garden is also a teaching facility, offering Master Gardener training to volunteers via classes, lecture series, workshops, and field trips. The garden also offers tours, plant sales, and various exhibits.

Hours are normally 8:30 A.M. to 4:30 P.M. Monday through Saturday and 1 P.M. to 5 P.M. on Sunday.

Slocum Water Gardens
1101 Cypress Gardens Boulevard
Winter Haven, FL 33884
941-293-7151
An aquatic nursery.
Open Monday through Friday 8 A.M. to 5 P.M. and Saturday 8 A.M. to 12 P.M.

Terry Park
3406 Palm Beach Boulevard
Ft. Myers, FL 33916
941-338-3232
Includes a demonstration butterfly garden.
Open Monday through Friday 8 A.M. to 5 P.M.

Check your phone book for county and state parks in your area for nature trails and pamphlets on native plants. Do not forget **Corkscrew Swamp Sanctuary**, an Audubon Society preserve near Naples; **Everglades National Park** in Dade County; and **Fakahatchee Strand** and **Big Cypress** in Collier County.

County Extension Services

The University of Florida's Extension Service has offices in every county. The addresses are:

Alachua County—District II
2800 NE 39 Avenue
Gainesville, FL 32609-2658
352-955-2402
fax: 352-334-0122

Baker County—District II
Route 3 Box 1074B
MacClenny, FL 32063-9640
904-259-3520
fax: 904-259-9034

Bay County—District I
324 West 6 Street
Panama City, FL 32401-2616
904-784-6105
fax: 904-784-6107

Bradford County—District II
2266 North Temple Avenue
Starke, FL 32091-1028
904-966-6224
fax: 904-964-9283

Brevard County—District III
3695 Lake Drive
Cocoa, FL 32926-8699
407-633-1702
fax: 407-633-1890

Broward County—District V
3245 College Avenue
Davie, FL 33314-7798
954-370-3725
fax: 954-370-3737

Calhoun County—District I
340 East Central Avenue
Blountstown, FL 32424-2206
904-674-8323
fax: 904-674-8353

Charlotte County—District V
6900 Florida Street
Punta Gorda, FL 33950-5799
941-639-6255
fax: 941-639-6719

Citrus County—District III
3600 South Florida Avenue
Inverness, FL 34450-7369
352-726-2141
fax: 352-344-4044

Clay County—District III
2463 State Road 16W
P.O. Box 278
Green Cove Springs, FL 32043-0278
904-284-6355
fax: 904-529-9776

Collier County—District V
14700 Immokalee Road
Naples, FL 33964-1468
941-353-4244
fax: 941-353-7127

Columbia County—District III
P.O. Box 1587
Lake City, FL 32056-1587
904-752-5384

Dade County—District V
18710 SW 288 Street
Homestead, FL 33030-2309
305-248-3311
fax: 305-246-2932

De Soto County—District IV
P.O. Box 310
Arcadia, FL 34265-0310
941-993-4846
fax: 941-993-4849

Dixie County—District II
P.O. Box 640
Cross City, FL 32628-1534
352-498-1237
fax: 352-498-1286

Duval County—District II
1010 North McDuff Avenue
Jacksonville, FL 32254-2083
904-387-8850
fax: 904-387-8902

Escambia County—District I
P.O. Box 7154
Pensacola, FL 32534-7154
904-477-0953
fax: 904-477-2893

Flagler County—District III
150 Sawgrass Road
Bunnell, FL 32110-0308
904-437-7464
fax: 904-586-2102

Franklin County—District I
33 Market Street Suite 305
Apalachicola, FL 32320-2310
904-653-9337
fax: 904-653-2261

Gadsden County—District I
2140 West Jefferson Street
Quincy, FL 32351-1905
904-627-6315

Gilchrist County—District II
P.O. Box 157
Trenton, FL 32693-0157
904-463-3174
fax: 904-463-3197

Glades County—District V
P.O. Box 549
Moore Haven, FL 33471-0549
941-946-0244
fax: 941-946-0629

Gulf County—District I
200 East 2 Street
P.O. Box 250
Wewahitchka, FL 32465-0250
904-639-3200
fax: 904-639-3201

Hamilton County—District II
P.O. Drawer K
Jasper, FL 32052-0691
904-792-1312
fax: 904-792-3205

Hardee County—District IV
507 Civic Center Drive
Wauchula, FL 33873-1288
941-773-2164
fax: 941-773-0958

Hendry County—District V
P.O. Box 68
Labelle, FL 33975-0068
941-674-4092
fax: 941-674-4098

Hernando County—District III
19490 Oliver Street
Brooksville, FL 34601-6538
352-754-4433
fax: 352-754-4489

Highlands County—District IV
4509 West George Boulevard
Sebring, FL 33872-5803
941-386-6540
fax: 941-386-6544

Hillsborough County—District IV
5339 County Road 579 South
Seffner, FL 32584-3334
813-744-5519
fax: 813-744-5776

Holmes County—District I
201 North Oklahoma Street
Bonifay, FL 32425-2295
904-547-1108

Indian River County—District IV
1028 20th Place, Suite D
Vero Beach, FL 32960-5360
561-770-5030
fax: 561-770-5148

Jackson County—District I
4487 Lafayette Street
Marianna, FL 32446-3412
904-482-9620
fax: 904-482-9287

Jefferson County—District II
275 North Mulberry
Monticello, FL 32344-2249
904-342-0187
fax: 904-342-0225

Lafayette County—District II
Route 3 Box 15
Mayo, FL 32066-1901
904-294-1279
fax: 904-294-2016

Lake County—District III
30205 SR 19
Tavares, FL 32778-4052
352-343-4101
fax: 352-343-2767

Lee County—District V
3406 Palm Beach Boulevard
Ft. Myers, FL 33916-3719
941-338-3232
fax: 941-338-3243

Leon County—District I
615 Paul Russell Road
Tallahassee, FL 32301-7099
904-487-3003
fax: 904-487-4817

Levy County—District II
P.O. Box 219
Bronson, FL 32621-0219
352-486-5131
fax: 352-486-1614

Liberty County—District I
P.O. Box 369
Bristol, FL 32321-0368
904-643-2229
fax: 904-643-5648

Madison County—District II
900 College Avenue
Madison, FL 32340-1426
904-973-4138

Manatee County—District IV
1303 17th Street West
Palmetto, FL 34221-2998
941-722-4524
fax: 941-742-5998

Marion County—District III
2232 NE Jacksonville Road
Ocala, FL 34470-3685
352-620-3440
fax: 352-368-5855

Martin County—District V
2614 SE Dixie Highway
Stuart, FL 33494-4007
561-288-5654
fax: 561-288-4354

Monroe County—District V
5100 College Road
Key West, FL 33040-4364
305-292-4501
fax: 305-292-4415

Nassau County—District II
P.O. Box 1550
Callahan, FL 32011-1550
904-879-1019
fax: 904-879-2097

Okaloosa County—District I
5479 Old Bethel Road
Crestview, FL 32536
904-689-5850
fax: 904-689-5727

Okeechobee County—District IV
458 Highway 98 North
Okeechobee, FL 34972-2303
941-763-6469
fax: 941-763-6745

Orange County—District III
2350 East Michigan Street
Orlando, FL 32806-4996
407-836-7570
fax: 407-836-7578

Osceola County—District III
1901 East Irlo Bronson Highway
Kissimmee, FL 34744-8947
407-846-4181
fax: 407-846-7286

Palm Beach County—District V
531 North Military Trail
West Palm Beach, FL 33415-1311
561-233-1712
fax: 561-233-1768

Pasco County—District IV
36702 State Road 52
Dade City, FL 33525-5198
352-521-4288
fax: 352-523-1921

Pinellas County—District IV
12175 125th Street North
Largo, FL 34644-3695
813-582-2100
fax: 813-582-2149

Polk County—District IV
Drawer HS03
P.O. Box 9005
Bartow, FL 33831-9005
941-533-0765
fax: 941-534-0001

Putnam County—District III
111 Yelvington Road Suite 1
East Palatka, FL 32131-8892
904-329-0318
fax: 904-329-1262

St. Johns County—District III
3125 Agriculture Center Drive
St. Augustine, FL 32092-0572
904-824-4564
fax: 904-829-5157

St. Lucie County—District IV
8400 Picos Road Suite 101
Ft. Pierce, FL 34945-3045
561-462-4660
fax: 561-462-1510

Santa Rosa County—District I
6051 Old Bagdad Highway Room 116
Milton, FL 32583-8944
904-623-3868
fax: 904-623-6151

Sarasota County—District IV
2900 Ringling Boulevard
Sarasota, FL 34237-5397
941-316-1000
fax: 941-316-1005

Seminole County—District III
250 West County Home Road
Sanford, FL 32773-6197
407-323-2500
fax: 407-330-9593

Sumpter County—District III
P.O. Box 218
Bushnell, FL 33513-0218
352-793-2728
fax: 352-793-0207

Suwannee County—District II
1302 11th Street SW
Live Oak, FL 32060-3696
904-362-2771
fax: 904-364-1698

Taylor County—District II
203 Forest Park Drive
Perry, FL 32347-6396
904-838-3508
fax: 904-838-3546

Union County—District III
25 NE 1st Street
Lake Butler, FL 32054-1701
904-496-2321
fax: 904-496-1111

Volusia County—District III
3100 East New York Avenue
DeLand, FL 32724-6497
904-822-5778
fax: 904-822-5767

Wakulla County—District I
P.O. Box 40
Crawfordville, FL 32326-0040
904-926-3931
fax: 904-926-8789

Walton County—District I
7320 North 9 Street Suite B
DeFuniak Springs, FL 32433-3804
904-892-8172
fax: 904-892-8175

Washington County—District I
1424 Jackson Avenue Suite A
Chipley, FL 32428-1615
904-638-6108
fax: 904-638-6181

Suppliers and Growers

Inclusion on this list does not imply an endorsement, and many fine sources are not included. Be aware that mailing addresses and phone numbers are subject to change.

General Suppliers

Brenda's Bloomers
2659 "G" Road
Loxahatchee, FL 33470
407-795-1734
fax: 407-795-0321
Hedge materials, birds-of-paradise, heliconias, ginger

Endangered Species
P.O. Box 1830
Tustin, CA 92681
714-544-9505
Rare bamboos, palms, exotic foliage

Kartuz Greenhouses
1408 Sunset Drive
Vista, CA 92083
619-941-3616
fax: 619-941-1123
Flowering plants; catalog $2.00

Logee's Greenhouses
141 North Street
Danielson, CT 06239
203-774-8038; 203-774-9932
Begonias, foliage plants, flowering plants, houseplants, conservatory plants, fragrant plants; catalog $3.00

Neon Palm Nursery
3525 Stony Point Road
Santa Rosa, CA 95407
707-585-8100
60 palm species, cycads, grasses, foliage plants, ferns, conifers

Reasoner's, Inc.
P.O. Box 1881
Oneco, FL 34264
941-756-1181
fax: 941-756-1882
Hibiscus, palms, ornamentals

Stoke's Tropicals
Box 9868
New Iberia, LA 70562
800-624-9706
Specializes in gingers, but also handles all kinds of tropicals

Sunshine Greenery
4740 Deer Run Road
St. Cloud, FL 34772
407-892-4893
Hibiscus, ixora, allamanda, croton, oleander, mandevilla

Tornello Nursery
115 12th Avenue SE
Ruskins, FL 33570
813-645-5445
fax: 813-645-4353
Bamboo

Tropical Paradise
5060 SW 76 Avenue
Davie, FL 33328
305-791-2029
fax: 305-791-7858
Large selection of flowering plants, shrubs, and trees; gingers and heliconias

Vista Nursery, Inc.
18100 SW 248 Street
Homestead, FL 33031
305-246-5200
fax: 305-246-5897
Woody ornamentals, mussaenda, foliage hibiscus, terrestrial orchids

Bromeliads

Alberts & Merkel
2210 South Federal Highway
Boyton Beach, FL 33435

Exotic Bromeliads
744 East Valencia Street
Lakeland, FL 33801

Holmes Nurseries
P.O. Box 17157
Tampa, FL 33607

Lee Moore
P.O. Box 504B
Kendall, FL 33156

Oak Hill Gardens
P.O. Box 25
Binni Road
Dundee, IL 60118

Earl Small
P.O. Box 11207
St. Petersburg, FL 33733

Orchids

B. F. Orchids
28100 SW 182 Street
Homestead, FL 33030

Fennel Orchid Jungle
26715 SW 157 Street
Homestead, FL 33031
Fort Caroline Orchids
13142 Fort Caroline Road
Jacksonville, FL 32225

J. E. M. Orchids
4996 NE Fourth Avenue
Boca Raton, FL 33430

Madcap Orchids
Route 29
Fort Myers, FL 33905

Oakhill Gardens
P.O. Box 25
Hinnie Road
Dundee, IL 60118

Orchid World International
11295 SW 94 Street
Miami, FL 33176

Orchids by Hausermann
2N 134 Addison Road
Villa Park, IL 60181

Gingers, Bananas

Banana Tree
715 Northhampton Street
Easton, PA 18842

Heliconia Haus
12691 SW 104 Street
Miami, FL 23186

Bulbs, Corms, Tubers

De Jager & Sons, Inc.
188 Asbury Street
South Hamilton, MA 01982

John Messelaar Bulb Company
P.O. Box 269
Ipswich, MA 01938

John Scheppers
63 Wall Street
New York, NY 10005

Water Lilies

Lilypons Water Garden
301 Lilypons Road
Brookshire, TX 77423

Perry's Water Gardens
919 Leatherman Gap Road
Franklin, NC 28734

Slocum Water Gardens
1101 Cypress Gardens Boulevard
Winter Haven, FL 33880

QUICK REFERENCE PLANT SELECTION

Plants with Showy Flowers

Acalypha
Allamanda
Alpinia
Anthurium
Bauhinia
Begonia
Bombax
Bougainvillea
Brownea
Brugmansia
Brunfelsia
Calliandra
Callistemon
Cassia
Clerodendrum
Clivia
Clusea
Costus
Delonix
Dichorisandra
Erythrina
Freycinetia
Hedychium
Heliconia
Hibiscus
Hymenocallis
Ixora
Jacaranda
Jasminum
Jatropha
Justicia
Kaempferia
Murraya
Mussaenda
Nerium
Petrea
Plumbago
Plumeria
Protea
Solandra
Spathodea
Stephanotis
Thunbergia
Tulbaghia
Zingiber

Plants with Red Flowers

Begonia
Bougainvillea
Canna
Lycoris
Hibiscus
Jatropha
Spathodea
Hedychium
Heliconia
Ixora

Plants with Yellow or Orange Flowers

Allamanda
Clivia
Canna
Lantana
Hibiscus
Begonia
Costus
Delonix
Heliconia
Plumeria

Plants with Blue, Violet, Purple Flowers

Bougainvillea
Hibiscus
Brunfelsia
Heliconia
Jacaranda
Thunbergia
Protea
Kaempferia
Petrea
Plumbago

Plants with White Flowers

Cleredendrum
Plumeria
Anthurium
Begonia
Heliconia
Hymenocallis
Jasminum
Nerium
Stephanotis
Zingiber

Shade Trees

Citrus spp.
Delonix
Erythrina
Eucalyptus
Ficus
Litchii
Mangifera
Palms
Persea

Foliage Plants

Aglaonema
Alocasia
Alpinia
Anthurium
Aphelandra
Begonia
Bromeliad
Calathea
Codiaeum
Cordyline
Cycads
Dracaena
Ferns
Ficus
Iresine
Monstera
Philodendron
Strobilanthes

Plants with Variegated Foliage

Aglaonema
Alpinia
Begonia
Bromeliad
Calathea
Codiaeum
Cordyline
Iresine

Fragrant Plants

Ardisia
Brunfelsia
Citrus spp.
Crinum
Eucalyptus
Hedychium
Jacaranda
Murraya
Nerium
Orchids (some)
Passiflora
Pittosporum
Plumeria
Spathiphyllum
Stephanotis

Selected Bibliography

Bell, C. Ritchie, and Byron J. Taylor. *Florida Wild Flowers and Roadside Plants*. Chapel Hill, North Carolina: Laurel Hill Press, 1982.

Berry, Fred, and W. John Kress. *Heliconia, an Identification Guide*. Washington and London: Smithsonian Institution Press, 1991.

Blombery, Alec, and Tony Todd. *Palms*. London, Sydney, Melbourne: Angus & Robertson, 1982.

Broschat, Timothy K., and Alan W. Meerow. *Betrock's Reference Guide to Florida Landscape Plants*. Cooper City, Florida: Betrock's Information Systems, Inc., 1991.

Courtright, Gordon. *Tropicals*. Portland, Oregon: Timber Press, 1988.

Kramer, Jack. *Orchids for the South*. Dallas, Texas: Taylor Publishing Co., 1994.

Lessard, W. O. *The Complete Book of Bananas*. Miami: W. O. Lessard, 1992.

Mathias, Mildred E., ed. *Flowering Plants in the Landscape*. Berkeley, California: University of California Press, 1982.

Meerow, Alan W. *Betrock's Guide to Landscape Palms*. Cooper City, Florida: Betrock Information Systems, Inc., 1992.

Menninger, Edwin A. *Flowering Trees of the World for Tropics and Warm Climates*. New York: Hearthside Press, Inc., 1962.

Morton, Julia F. *500 Plants of South Florida*. Miami: E. A. Seemann Publishing, Inc., 1974.

Myers, Ronald L., and John J. Ewel, eds. *Ecosystems of Florida*. Orlando: University of Central Florida Press, 1990.

Neal, Marie. *In Gardens of Hawaii*. Honolulu: Bishop Museum Press, 1965.

Shuttleworth, Floyd S., Herbert S. Zim, and Gordon W. Dillon. *Orchids*. Racine, Wisconsin: Golden Press, 1986.

Smiley, Nixon. *Florida Gardening Month by Month*. 3rd ed. Coral Gables, Florida: University of Miami Press, 1986.

Stresau, Frederic B. *Florida My Eden*. Port Solerno, Florida: Florida Classics Library, 1986.

Stevenson, George B. *Palms of South Florida*. Miami: Fairchild Tropical Garden, 1974.

———— *Trees of the Everglades National Park and the Florida Keys*. Miami: Banyan Books, 1969.

Tasker, Georgia. *Wild Things, the Return of Native Plants*. Winter Park, Florida: The Florida Native Plant Society, 1984.

Watkins, John V., and Thomas J. Sheehan. *Florida Landscape Plants, Native and Exotic*. Revised Edition. Gainesville, Florida: The University Presses of Florida, 1975.

Index

Numbers in *italics* indicate a photograph

Q–R

S

T

U–V

W–Z